VOLUME TWO HUNDRED AND FIVE

ADVANCES IN IMAGING AND ELECTRON PHYSICS

EDITOR-IN-CHIEF

VOLUME TWO HUNDRED AND FIVE

ADVANCES IN IMAGING AND ELECTRON PHYSICS

Edited by

PETER W. HAWKES
CEMES-CNRS
Toulouse, France

Cover photo credit:
The cover picture is taken from Fig. 10A of the chapter by Erich Plies (p. 186).

Academic Press is an imprint of Elsevier
125 London Wall, London EC2Y 5AS, United Kingdom
525 B Street, Suite 1800, San Diego, CA 92101-4495, United States
50 Hampshire Street, 5th Floor, Cambridge, MA 02139, United States
The Boulevard, Langford Lane, Kidlington, Oxford OX5 1GB, United Kingdom

ISBN: 978-0-12-815217-1
ISSN: 1076-5670

Publisher: Zoe Kruze
Acquisition Editor: Jason Mitchell
Editorial Project Manager: Shellie Bryant
Production Project Manager: Divya Krishnakumar
Designer: Alan Studholme

Typeset by VTeX

CONTENTS

CONTRIBUTORS

Erich Plies
Formerly Forschungslaboratorien, Siemens AG, München and University of Tübingen, Germany

Dirk van Delft
Rijksmuseum Boerhaave, Leiden, The Netherlands

John van Gorkom

Ton van Helvoort
Independent scholar

PREFACE

This volume contains two substantial contributions, the first of which consists of two lightly edited chapters from the draft dissertation of the late John van Gorkom. The background to this is presented by Dirk van Delft, director of the Rijksmuseum Boerhaave in Leiden, and Ton van Helvoort, a freelance writer, at the end of van Gorkom's text. After this, a further article from *Advances in Optical and Electron Microscopy* is reproduced. This is a scholarly account by Erich Plies of the electron optics of low-voltage electron beam testing and inspection.

The subject of van Gorkom's chapter is the early development of the electron microscope, from Busch's publications of 1926 and 1927 to the first experimental work in the Technische Hochschule and the AEG Research Laboratory in Berlin, with occasional glances at de Broglie's work, published between 1923 and 1925. The various patents applied for by Knoll, Rüdenberg, Ruska, and the AEG team are examined minutely and the findings of Lin Qing (not hitherto available in English) are recapitulated in detail and sometimes criticized. As well as the more formal material, van Gorkom includes many human touches and comments on the motives of the various actors. The argument is not based solely on published sources; in 1985, van Gorkom and a friend visited Ernst Ruska and recorded their conversations with him. The essence of these interviews is woven into the story that van Gorkom tells.

The text has not been modified, apart from some essential improvements in grammar and syntax, which van Gorkom would doubtless have made before submitting the dissertation. In particular, I have not altered English translations of material from Ruska's historical article in the *Acta Leopoldina* even though an English translation is available.

There is one point that I should like to have raised with van Gorkom. He suggests that the letter sent by Max Steenbeck to Max Knoll in 1961 in response to a request for information about his visit to Knoll and Ruska's laboratory in 1931 should not be accorded too much importance as Steenbeck was in East Germany at the time and may not have felt able to write freely. I find this suggestion (for which there is no documentary evidence) unlikely. In the 1960s, I corresponded with Prof. Dr. Johannes Picht, who was in Potsdam (in the German Democratic Republic) with no sense of restraint or caution. He lent me various unpublished documents and from the tone of his letters it was clear that he was not worried about censorship. I can add a small detail concerning the Knoll–Steenbeck correspondence (reproduced in *The Beginnings of Electron Microscopy*, Supplement 16 to the present serial). In a letter dated 22 February 1960, Professor Erwin Weise of the University of Illinois in Urbana, wrote to ask Knoll to clarify the early history of the electron microscope:

> *Lieber Herr Knoll!*
> *Ich bringe hier zusammen mit einem älteren Kollegen von der Chemie eine Encyclopedia of Chemistry heraus; er ist der "editor" und hat mich um Mitarbeit gebeten. In dem Buch soll auch eine Menge über das Elektronenmikroskop gesagt werden. Leider ist aber das historische Bild sehr unklar. So hat uns Professor Rüdenberg einen langen Aufsatz geschickt in dem er behauptet, Alleinerfinder zu sein. Ihr Name wird so am Rande erwähnt. Sie hätten zwar über Fokussierungsbedingungen an Elektronenstrahlen gearbeitet aber niemals an eine Bilderzeugung gedacht. Dies hätte er allein zuerst getan und in einem Französchen Patent im Jahre 1932, mit der Priorität von 1931, veröffentlicht. Sein Deutsches Patent von 1931 sei durch die Naziregierung zurückgehalten und erst 1954 veröffentlicht worden. Inzwischen hätten dann andere Erfinder, wie Knoll, Ruska, Ardenne usw, unverdiente Anerkennung gefunden. Wir wollen nun unbedingt gerecht sein, und mein Kollege, ein Professor George L. Clark, dessen X-ray Buch Sie vielleicht kennen, hat mich gebeten bei Ihnen anzufragen, wie die Geschichte nach Ihrer Meining war. Ich wäre Ihnen sehr dankbar, wenn Sie uns mit Hinweisen auf früheste Veröffentlichungen und auch mit Auskunft anderer Art helfen könnten.*

The Encyclopedia mentioned should certainly be the *Encyclopedia of Microscopy*, as the *Encyclopedia of Chemistry* had already been published in 1957 and its second edition did not appear until 1966. The *Encyclopedia of Microscopy* was published in 1961, shortly after the *Encyclopedia of Spectroscopy* to which it formed a pendant. It was thus this letter that led Knoll to write to Steenbeck and incidentally led to the oft-repeated (if implausible) assertion that Rüdenberg intended his manuscript for the *Encyclopedia of Chemistry*.

Readers wishing to pursue the subject may be interested in a small book by Götz von Borries, a cousin of the Bodo von Borries who plays a large part in the history. *Bodo von Borries und das Elektronenmikroskop, Erfindung und Entwicklung* (Fouqué Literaturverlag, Egelsbach bei Frankfurt am Main, 2001) retraces the story from a somewhat different standpoint. A very full study by Falk Müller will soon be published.

The title of the article by Erich Plies does not do justice to the breadth of the material covered. There are sections on methods of calculating the magnetic and electrostatic fields, paraxial optics and aberrations, including the parasitic aberrations of the complicated devices described, and

Coulomb interaction and charging. Although there have been many developments of the subject since 1994, when the article first appeared, it remains an excellent guide to the topic. I am sure that it will enjoy a second life now that it is readily available in ScienceDirect as well as in book form.

Peter W. Hawkes

CHAPTER ONE

The Early Electron Microscopes: A Critical Study*

John van Gorkom✠, with the cooperation of Dirk van Delft and Ton van Helvoort

Contents

* This article represents two chapters from the draft doctoral thesis of the late John van Gorkom. See Appendix B. E-mail for correspondence: hawkes@cemes.fr.
✠ Deceased.

Advances in Imaging and Electron Physics, Volume 205
ISSN 1076-5670
https://doi.org/10.1016/bs.aiep.2018.01.001

PART 1. CONCEIVING THE IDEA

Just four years elapsed between Hans Busch's publication of his lens theory (Busch, 1926, 1927) and the construction in 1931 of the very first device that could be called the primordial electron microscope. In these years, the concept of electron optics had caught the imagination of many according to Dennis Gabor, who told a British audience in 1942:

> *"Busch's paper was more than an eye-opener; it was almost like a spark in an explosive mixture. In 1927 the situation in physics was such, that nothing more than the words 'electron lens' were needed to start a real burst of creative activity"* (Gabor, 1942).

And he added that for experimental physicists,

> *"the words 'electron lens' were enough (...) to ask themselves: 'What can we do with electron lenses? Can we make electron microscopes, electron telescopes, etc.?"*

Whether Gabor's representation of the facts is altogether correct remains to be seen as will become clear later where the conception of the idea of an electron microscope in the period spanning approximately 1928 to early 1932 will be discussed extensively. The structure of this part is based on the existing priority claims with respect to the invention of the instrument. Since Ernst Ruska received the 1986 Nobel Prize in Physics "for the design of the first electron microscope", as the Nobel Committee formulated it, it seems fair to start with the story of the electrotechnical student Ruska and his tutor Max Knoll at the Berlin *Technische Hochschule*. The second section is dedicated to the claims by Ernst Brüche, who worked for AEG—the German sister company of General Electric. The third section deals with the patents on the electron microscope that were filed by Reinhold Rüdenberg on behalf of AEG's competitor Siemens. In the fourth part attention is paid to Dennis Gabor's statement that the later nuclear physicist Léo Szilárd had been the first to come up with the idea. In the fifth part, the situation in Great Britain and the USA is briefly mentioned for the sake of completeness. Finally, in the last part the major conclusions from the five preceding sections will be brought together and merged into one.

Much attention will also be given to the motives of those immediately involved, as this relates to the validity of certain claims and in particular to the question of what engineers wished to achieve when they developed the transmission electron microscope.

1. ERNST RUSKA, MAX KNOLL AND THE TECHNISCHE HOCHSCHULE

1.1 Background

The fact that the first prototype of a transmission electron microscope was constructed at the *Hochspannungslaboratorium* (High Tension Laboratory) of the Technische Hochschule in Berlin is undisputed. The laboratory was run by Adolf Matthias, the director of the *Studiengesellschaft für Höchstspannungsanlagen* (Research Society for Very High Tension Facilities), which had been founded in 1921 by the *Vereinigung der Elektrizitätswerke e. V.* (Association of Electric Companies). In 1925 Matthias had been made *Honorarprofessor* (unsalaried professor) of High Tension Technology and Electrical Facilities at the Technische Hochschule, an appointment that was changed into a full professorship a year later (Grünewald, 1976; Freundlich, 1994). The first thing Matthias did in 1926 was to set up the High Tension Lab in a roofed-over inner court of the Hochschule's Electrotechnical Institute in which he placed "a huge 3-stage, air-cooled, 50 Hz, one megavolt transformer", as reported by Martin Freundlich, one of Matthias' first PhD-students at the time (Freundlich, 1994). As the institute's building survived World War II in some miraculous way, it can still be found at *Strasse des 17. Juni* (June 17th Street)—then called *Berliner Strasse* at that particular stretch—at the south eastern side of today's *Ernst-Reuter-Platz*—known at the time as *Knie*—in the area called Charlottenburg.

One of the problems that Matthias had to address was the incidence of electrical surges in the electricity network. During thunderstorms strikes of lightning may induce high voltage shock waves in power lines. These shock waves—which can also be caused by other events such as short circuiting—can destroy all sorts of equipment when they shoot through the network. To learn more about the problem, Matthias wanted to develop a measuring instrument that would be robust enough to make reliable recordings of these waves (Ruska, 1979; Grünewald, 1976; Freundlich, 1994). For this purpose he chose to develop a high-speed oscilloscope.

At that moment, someone else in the Electrotechnical Institute was already doing research on the oscilloscope. It happened to be Dennis Gabor, who had started in 1924 as a student of professor Ernst Orlich on a PhD to improve the instrument (Keith, 2008). One of the changes that Gabor had made was to partly encase the concentrating coil (Ruska, 1979; Grünewald, 1976; Freundlich, 1994) that was generally employed ever since its introduction by Wiechert. Gabor had noticed that the large magnetic field of the coil interfered with the deflectors that steer the cathode ray; the rotational effect of the field would cause a distortion of the oscillograms. To shield off this field, Gabor applied an iron cover to the outside of the coil. In later years it has been suggested that he had applied the iron shell to improve the focusing properties of the coil (Ramsauer, 1941) but such was not the case. As Gabor stressed himself thirty years later, no one understood the concentrating properties of these coils yet (Gabor, 1957).

After the creation of the new High Tension Lab at the Electrotechnical Institute, Gabor joined Matthias who became his supervisor until he completed his PhD in 1927. Matthias decided that Gabor's fruitful work should be continued and in 1928 he appointed his assistant Dr. Max Knoll as head of a new project team, which became informally known as the *Knoll Gruppe*. Soon, Knoll was looking after PhD-students Henning Knoblauch, Gerhard Lubszynski and afore-mentioned Martin Freundlich. In his last class before the 1928 summer break, Professor Matthias also invited Master students to join the oscilloscope project. There had only been one who responded: the 21 year old Ernst Ruska (Ruska, 1970a). The next spring, on 1 April 1929, the group was finally joined by the PhD-student Bodo von Borries, one of the other main characters of this history (Freundlich, 1994; Grünewald, 1976; von Borries, 1991).

On many occasions it has been recounted that Knoll was an excellent team builder. Upon receiving the prestigious *Paul-Ehrlich- und Ludwig-Darmstaedter-Preis*, Ernst Ruska told his audience in 1970:

> *"Long before the concept of "teamwork" was introduced (...) Max Knoll went to great lengths to train us to cooperate well" (Ruska, 1970a, 1974).*

And in the 1985 interview that Ruska gave to Bart de Haas and me [JvG], he told us:

> *"Of course we had a coffee break each day, and there we talked about our common interests. That was a very fruitful exchange of ideas, in which everyone expressed his views frankly, and [so] we helped each other" (van Gorkom & de Haas, 1985).*

Freundlich (1994), finally, completed this image in 1994, writing:

> *"Knoll was the ideal leader for advising and encouraging the young students. During the afternoon coffee breaks, the successes and setbacks of the experiments were discussed."*

1.2 Ernst Ruska's First Assignment in 1928–1929

In this stimulating environment Ruska worked on his first assignment from 1 November 1928 until 10 May 1929. The *Studienarbeit* (student research project) that he completed during these six months was titled "On a method to calculate the cathode ray oscilloscope on the basis of the experimentally found dependence of the writing dot's diameter on the position of the concentrating coil" (Ruska, 1929). In other words, it had been Ruska's assignment to find a way to calculate the best position of the concentrating coil with respect to the diameter of the spot that is created on the oscilloscope's viewing screen by the cathode ray. This project fitted well in Matthias's endeavour to develop an industrial oscilloscope that would be able to record high voltage shock waves. These waves flash by in one hundred millionth of a second. As a result the writing speed of the dot on the oscilloscope's screen was 3,000,000 m/s. With such speeds you need an extremely bright and sharp dot to be able to see any trail left on the screen after the invisibly fast deflection of the beam. So basically it meant that it had been Ruska's job to make the writing dot as small and intense as possible. He was fortunate, therefore, to join in just after Hans Busch had published his lens theory (Busch, 1926, 1927).

First of all, however, it was necessary to prove the theory experimentally, as Busch himself had but very poor experimental data at his disposal, which he had collected 16 years before. In the book that Ruska published in 1979 as well as in his 1986 Nobel Prize Lecture, Ruska even wondered whether this lack of reliable data might have made Busch especially hesitant to go into any speculations about novel electron optical applications himself (Ruska, 1979, 1986). Whatever the case was, in 1929 Ruska managed to show that Busch's theory was perfectly sound. For this purpose he used a heavily modified, expendable oscilloscope which was stripped of its deflector plates. Not surprisingly for a high tension lab, the device operated under high tension, ranging from 30 to 70 kilovolts. The high electron speeds that result from such tensions were known to reduce the ionisation of residual gas in the tube (Ruska & Knoll, 1931; Knoll & Ruska, 1932a, 1932b). The creation of ions is unwanted since they will interfere with the electron beam and disturb the optical behaviour which was the very subject of the Studienarbeit. From the experiments with this device Ruska was able to conclude that the discrepancy between his experimental outcomes and the expected values was only 5 percent (Ruska, 1929; two years later, the same conclusion was drawn in Ruska & Knoll, 1931). This finding provided him with a powerful tool to determine the right position for the concentrating coil.

The question that is of most interest here is to what extent Ruska was aware that he had entered the completely new area of geometric electron optics the very moment that he had proven the lens formula to be valid. And if he did realise this, whether he was aware of its potential. Since we already saw that there was a daily exchange of ideas with his colleagues, the question actually extends to the whole Knoll Group. This matter has been addressed before, in 1995, by Qing Lin who examined Ruska's Studienarbeit closely (Lin, 1995). The point of departure for him was a quote from a press release by the Royal Swedish Academy of Sciences which says about the electron microscope:

> *"Its development began with work carried out by Ruska as a young student at the Berlin Technical University at the end of the 1920's" (Royal Academy of Sciences, 1986).*

Wondering whether this was a correct representation of facts, Lin describes the experimental setup that Ruska used to prove the correctness of Busch's lens theory and observes:

> *"You might well say that this experimental setup resembles the 'optical bench' in light optics. Likewise, the performed measurements have an explicit optical character."*

This brings him to the conclusion that:

> *"By checking Busch's formula the [Knoll] Group had established a connection with geometric electron optics."*

Subsequently, however, Lin insists that this does not mean that Ruska had any awareness of what these new insights might imply. The Knoll Group occupied itself with the dimensioning of oscilloscopes, and any suggestion that they were realising where this new optics might take them, has to come from retrospective interpretation, Lin argues. Since the issue of retrospection is a recurrent theme in this article, I wish to stress that Lin does not apply a valid argument. Our inability to know the future does not exclude the ability to have visions, dreams or fantasies. There is no a priori ground to exclude the possibility that a member of the Knoll Group might have had a vision or fantasy. The real truth is that we simply do not know.

As a matter of fact, in this respect Lin either overlooked or dismissed a very relevant aspect of Ruska's study, which is the iron casing of the concentrating coil. The iron casing of an electron lens will reduce the axial length of the magnetic field to approximately the size of the small concentric gap at the inside of the coil. This reduction of the length of the field is vital to constructing a lens with short focal distance, while a lens with short focal distance is vital to the concept of microscopy. In short, the casings of concentrating coils are essential to electron microscopes. As mentioned above, Gabor had already applied an iron cover to the concentrating coil of the oscilloscope that he had left behind in the High Tension Lab after completing his PhD in 1927. His cover did not extend into the inside of the coil, since Gabor's sole intention had been to keep the magnetic field away from the deflector plates. Nevertheless it does mean that the idea of encasing the coil was not unfamiliar to the members of the Knoll Group who were supposed to have picked up the development where Gabor had left it.

From a historical point of view it is of great interest to know at which moment the first signs appear that Ruska realised that an iron casing of the coil could also be used to reduce its focal length. In three publications from 1957, 1974 and 1987—the latter being his Nobel Lecture—Ruska is clearly suggesting that he did so in 1928–29 during his Studienarbeit, and moreover, that he actively applied such a casing then (Ruska, 1957, 1974, 1987). Quite strangely though, this turns out not to be the case when examining his Studienarbeit. In the theoretical part of the study, he dedicated 24 pages to the calculations of concentrating coils, of which only the last page treated the subject of encasing. The reason for him to address the issue appears to have been the belief by some at the time that a casing would enhance the further reduction of the writing dot's size. In response to this he writes:

> *"Without wanting to anticipate on the experimental verification of this question, from the point of view of the Busch theory on cathode ray beams one can only say that there exists no theoretical ground for a reduction of the dot" (Ruska, 1929).*

As such, this is a valid statement, but it clearly reveals that Ruska had not made any attempts yet to perform such an experimental verification. He does continue to acknowledge, however, that a casing would compress and intensify the magnetic field. As a result there would be "a slight reduction" of the current needed to create a magnetic field that will achieve the same concentrating effect. There is no mention of a reduced focal length here, even though such a reduction was implicit in the formulas on which Ruska's statements were based.

Does this mean that Ruska deliberately tried to bend the historical facts? At least on one occasion he was very explicit that indeed he had been using coils *without* casing at the time, which was in 1970 when he gave his acceptance speech after receiving the Paul-Ehrlich- und Ludwig-Darmstaedter-Preis (Ruska, 1970a). And in the book that he published in 1979 he only claims to have understood the advantages of a compressed field, but does not say that he actually experimented with it during his Studienarbeit (Ruska, 1979). As we will see in the following paragraphs, it looks as if Ruska did make these experiments some time later, after obtaining his engineering degree, so maybe in later years he simply tended to confuse the order of events.

Although Lin has not taken this whole issue of the casing into consideration, it was already quite clear to him that the Studienarbeit does not contain any proof that Ruska had grasped the idea of true electron optics at that moment. Indeed, since Ruska's immediate assignment related to the improvement of the oscilloscope, in his report you will only find references to the concentration of the electron beam in order to reduce the diameter of the writing dot on the screen. Just two optical terms can be found and both of them had already been used by Hans Busch himself. One term is "focal distance", which he puts between quotation marks when introducing it (Ruska, 1929). The use of quotation marks here is not very remarkable, however, as at that stage the validity of Busch's lens theory had yet to be proven, including the concept of the focal distance of an electromagnetic coil. The other term is the "spherical aberration", which Ruska mentions in order to discuss the sharpness of the writing dot's edge (Ruska, 1929). Besides this, it is also worth noting that there is not the slightest reference to de Broglie's wave–particle duality, which is so important to the intuitive version of the electron microscope's history.

Although Lin does not say so literally, his examination of the Studienarbeit leaves the strong impression that the Nobel Committee had no ground to suggest that the development of the electron microscope started at the end of the 1920s. To some extent, though, such a judgment must be based on semantic if not philosophical arguments. After all, is it imperative that a scientific development is a fully conscious experience right from the beginning? It is doubtful whether this is ever the case when people enter completely new fields of research. Consciously or not, at certain moments scientists decide to move into directions which later prove to have been choices of great significance. In fact, if it is the wish to avoid a retrospective interpretation of history, one might even want to exclude the whole question whether historical characters had a clear vision, since at that stage any vision is supposed to be just a wild guess.

1.3 Ernst Ruska's Second Assignment in 1930

There are no records available revealing what Ruska did in the year after he finished his project on 10 May 1929. He had been studying in Munich for two years from 1926 till early 1928, and after his pre-examination there, he went to live with his parents again who had just moved to Berlin after his father Julius had been appointed Director of the newly founded Research Institute for the History of the Natural Sciences (Ruska, 1987; Lambert & Mulvey, 1996; Kraus, 1938). This means that in May 1929 Ernst Ruska was finishing in Berlin the third year of his study, and will have had more classes to attend in the following year.

In order to obtain his degree as *Diplom-Ingenieur*—a degree in engineering comparable to a M.Sc.—Ruska had to make a *Diplomaufgabe* (diploma assignment). According to the cover page of his assignment paper, Ruska started on his Diplomaufgabe on 18 July 1930, more than a year after completing the Studienarbeit. He devoted it to electron optics again—this time to the "Study of electrostatic converging devices as alternatives to magnetic concentrating coils in cathode ray oscilloscopes," as his paper is called (Ruska, 1930). The theoretical possibility that electrostatic fields will act like lenses, had also been given by Hans Busch in 1927. Electrostatic fields are generated by static electricity, an everyday phenomena that simply means that there are more electrons at a certain place than in its vicinity. From a technical point of view, electrostatic lenses are much more attractive than electromagnetic ones, because they are far simpler to make and far easier to keep steady on a chosen tension. Naturally, 'ease' is a powerful argument to engineers, who always have to mind about the practical feasibility of their designs.

Ruska's tutor was again Max Knoll, who in the mean time had proven that his interest in electrostatic lenses was so strong that he had even filed the very first patent ever on one on 10 November 1929—half a year before Ruska started on his new assignment. Knoll's lens consisted of a sandwich of three sheets with a hole in the middle of that sandwich. The two sheets at the outside would have the same charge, which would differ from the sheet in between. As a result, an electron beam travelling through the hole in the three sheets will experience a radially symmetrical electrostatic field (Knoll, 1929); Lin (1995) argues that Knoll's patent merely describes a concentrating device, since Knoll also proposes an arrangement that is composed of five sheets with opposite charges, as an alternative to the arrangement with three sheets.

Considering the existence of this patent, something irrational happened when Ruska started on his Diplomaufgabe, as he has pointed out himself in his book as well as his Nobel lecture. On 8 December 1986 he described to his audience in Stockholm his talks with Knoll on the latter's design of an electrostatic lens:

> *"We discussed the shape of the electric field between these electrodes [sheets, JvG], and I suggested that because of the mirror-like symmetry of the electrostatic field of the electrodes on either side of the lens centre, a concentrating effect (...) could not take place. I only had the field geometry in mind then. But this conclusion was wrong. I overlooked that as a consequence of the considerably varying electron velocity on passage through such a field arrangement, a concentration of the divergent electron bundle must, in fact, occur. Knoll did not notice this error either. (...)"*

Because of this error of judgment, Ruska did not implement Knoll's patented design, but constructed a so-called *Kugelkondensator* (ball-shaped condenser lens). This condenser lens consisted of a solid metal sphere of 1.5 cm diameter, centred inside another bigger one with an inner diameter of 6 cm. These two had different electrical charges in order to create an electrostatic field. As an electron beam had to pass through it, both spheres had 6-mm-wide holes, and in order to preserve the shape of the field, the holes in the inner sphere were covered with very fine metal mesh screens. Not surprisingly, this was not an ideal arrangement since the electron beam had to pass through the mesh screens which would diminish the intensity of the cathode ray considerably, partly by absorption and partly by scattering. As it was the sole purpose of the condenser lens to *increase* the intensity of the cathode rays, this was the most adverse effect you can think of.

There was, nevertheless, also something good in the bad results of Ruska's Diplomaufgabe, as he told his listeners in Sweden in 1986:

> *"As a consequence of my false reasoning and the experimental disappointment I decided to continue with the magnetic lens. I only report this in so much detail to show that occasionally it can be more a matter of luck than of superior intellectual vigor to find a better—or perhaps the only acceptable way."*

As history has proven, electrostatic lenses are not very well suited for electron microscopy, despite the numerous efforts that were made in later years to develop electrostatic microscopes.

Apart from this accidental learning experience, there are two other observations in Ruska's Diplomaufgabe that are certainly worth mentioning when it comes to significant outcomes in the view of later developments. Although the tone of voice of Ruska's report on this second assignment is just as conservative as his first one when it comes to optical observations, he does mention that:

> *"When employing an unconcentrated beam, often the image of the mesh structure appeared on the screen in a magnification of 50 to 100 times, and actually this image was considerably larger than the unconcentrated dot would normally be. This phenomenon was not reproducible at will, and only occurred with certain voltages and currents of the discharge tube [electron gun, JvG] when the inner ball was moved perpendicular to the axis of the beam" (Ruska, 1930).*

If you consider that practically speaking geometrical electron optics did not exist yet, these magnified images must have been quite intriguing in the second half of 1930.

Subsequently, he described at the very end of his paper an effort to replace the mesh screens by very thin aluminium foil that would allow 80 percent of the electrons to pass through, in order to find out if such a foil would work better. Ruska had obtained this percentage from the work of Philipp Lenard (Lenard, 1918). The effect of the foil was the opposite, however, and Ruska writes as last words of his report:

> *"The foil simply scatters the electron beam diffusely, when it passes through it, just as a pane of frosted glass scatters the light that passes through it."*

Scattering is of great importance in electron microscopy and the awareness of that importance is one of the major themes of the history of the electron microscope.

As Lin already pointed out in his study, the final historical significance of Ruska's attempt to construct an electrostatic lens is the strengthening of his ties to the nascent field of electron optical research after his first introduction to it more than a year before. To what extent this was consciously experienced by him is hard to judge on the basis of available sources. Ruska himself said in 1957:

> *"The goal of our group was a technical one from the beginning on. Because all of us younger colleagues were engineers, we were interested in the constructional design of the cathode-ray oscilloscope in the first place. Nevertheless, we soon occupied ourselves as well with electron optical lines of thinking. These developed as a consequence of our calculations on the focal distances of magnetic electron lenses, and*

the successful attempts to image defined cross-sections of the electron beam. This clarified our views on the optical character of electron rays and on the possibilities to build electron lenses and electron optical devices" (Ruska, 1957).

As such, this all sounds quite plausible. They were young and eager, they knew the papers of Hans Busch, they had made it a habit to exchange their views during coffee breaks and so they started by and by to recognize the true significance of electron optics. Whether it was realistic, though, to imagine at that stage an optical device like an electron microscope is far less obvious.

1.4 Construction of the Primordial Electron Microscope

Ruska turned in his thesis on 23 December 1930, two days before his 24th birthday. Shortly afterwards he received his degree in engineering. At that moment, it would have been perfectly normal to start looking for a paid job, but there was a world-wide economic recession in progress, which had started with the historic Wall Street Crash in October 1929. This ill wind, however, blew him luck as well. As he was living at home, he was allowed by his parents to take up an unpaid position as PhD-student at the High Tension Lab of Professor Matthias.

Without warning, the next thing Ernst Ruska did was to build the very first prototype of an electron microscope with two concentrating coils, enabling him to make so-called two-stage images of an object. It is not very clear, where this abrupt development exactly came from, but most likely it will have been the result of that daily exchange of ideas during the repeatedly mentioned coffee breaks. When asked for a comment during the interview that we had with Ruska in 1985, he told us plainly:

"I did not know yet what I exactly wanted to do for my PhD. Therefore I first tried the two-stage image. And when this succeeded, we discussed how we might make lenses with shorter focal distances, as with such lenses one can achieve high magnifications without a very large length of the apparatus" (van Gorkom & de Haas, 1985).

The person with whom he discussed these plans was Max Knoll, who had become his supervisor after his graduation. Ruska succeeded in making the two-stage images on 7 April 1931, according to his laboratory notebook. The drawing in his notebook of the apparatus that he had constructed, is dated 9 March 1931 (Ruska, 1979). It shows a device that is standing upright, with the viewing screen at the bottom, the high voltage

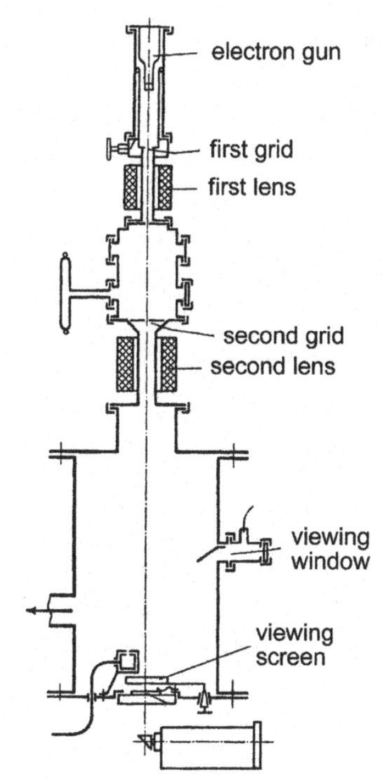

Figure 1 Cross section of experimental apparatus. Based on Knoll and Ruska (1932b).

electron gun at a safe place up in the air,[1] and two electromagnetic coils, which still had no casing (Ruska, 1934a). The photographs of the first two-stage electron images have become historical, though for a layman it will be difficult to recognise their importance. As often in history, the first results were very sketchy with an unimpressive magnification of 17 times in total for the two coils. The significance, however, was the fact that Ruska had proven that you can use a second coil to magnify the image obtained with the first one.

He had shown this by using two grids as objects. He had placed the first grid before the first coil and the second grid before the second coil, as shown in Fig. 1. The first coil would project the image of the first grid on

[1] It was more common to operate cathode-ray tubes in a horizontal position, as this makes it easier to shift the concentrating coil a bit to the left or right from outside of the tube. The earlier devices that Ruska experimented with in his Studienarbeit and Diplomaufgabe also stood upright. So did Gabor's earlier experimental oscilloscope, as can be seen in Mulvey (1962). The main reason was safety.

top of the second grid. This means that a virtual object (the image of the first grid) was added to a real object (the second grid). Subsequently, the second coil produced a combined magnification of the virtual grid plus the real grid—that is, a two-step magnification of the first grid, superimposed on a one-step magnification of the second grid. The fact that this was possible, clearly demonstrated the purely optical behaviour of electron rays. It definitely meant that it was permissible to call these electromagnetic coils electron lenses.

Is it fair, then, to call this device that Ruska had constructed, the first electron microscope ever? Dennis Gabor wrote in 1957:

> *"Maybe it is not so much an electron microscope as it is more a kind of electron optical bench."*

Tom Mulvey explained to an audience in 1961:

> *"This apparatus, the first compound electron microscope, is probably more accurately described as an electron-optical bench since it was designed primarily for making accurate measurements of magnification rather than for looking at specimens" (Mulvey, 1962).*

Heinrich Grünewald speaks in 1976 of "an electron optical bench as first electron microscope." And finally there is Qing Lin, who in 1995 appears to sympathise with Gabor's qualification. Although all three quotes show a preference for calling it an "electron-optical bench", no-one denies that it was the primordial version of the electron microscope.

Again, it is also a matter of perspective. If you focus on its original purpose, you would have to agree with Mulvey. If you focus on its performance, it is difficult to call it an electron microscope, for the simple reason that its ability to magnify hardly exceeds that of a magnifying glass. When you look at the basic features, however, everything is there: the electron gun, several electromagnetic lenses, an object, a projection screen and the ability to produce two-step magnifications. To be frank, the final judgement is rather arbitrary.

What has to be taken into consideration as well, however, is Ruska's own perception. During the 1985 interview, we asked Ruska several times about his insights and motives when he made his early experiments. Of course, personal memories are subjective and certainly less reliable after so many years, but as far as we could judge, Ruska was nevertheless very accurate and straightforward all the time. As we just saw already, Ruska had clearly stated that first he wanted to see whether the two-stage image would succeed, before he would move on. This impression of a preconceived plan also arose when he answered our question what his goal was at the time. At that occasion he said that:

> *"It was a conscious, purely optical line of thinking: whether one can reproduce with electrons the optical laws that are important for a two-stage image by a microscope. There was already the idea of a microscope in the background, to make high magnifications of objects transilluminated by electrons" (van Gorkom & de Haas, 1985).*

Two days later we asked Ruska at which moment he started to believe that electron microscopy was feasible and he answered:

> *"That must have been at the beginning of 1931. (...) Two years after my first work on those lenses and the test of this theory of Busch, I had to do my exams. When I could not find a job after my graduation, my parents said that I should do my PhD. I then immediately started—certainly after talks with Knoll—to construct an apparatus with two lenses, to try whether you can further magnify an image with a second lens. That is a microscope, isn't it? (...) From my upbringing I knew what a microscope is, and how important a microscope is, and my brother (...) knew as well. And of course we have frequently discussed these matters together."*

Ruska's brother Helmut was two years younger and already in their childhood they tended to team up. By the time that Ernst Ruska made his first experiments at the Technische Hochschule, Helmut was studying medicine, and therefore he was certainly well acquainted with microscopy and very well aware of its significance.

Their father Julius was a typically old school nineteenth century German scholar. He had studied philosophy, mathematics and natural science, after which he had specialised in the history of the natural sciences in Islam. For that reason he had also acquired a great knowledge of Middle-Eastern languages during his life (Kraus, 1938). Father Julius used to work at home, which did not always go very well together with having children, as Ernst told, for instance, in the Nobel lecture that he gave on 8 December 1986:

> *"In the second floor of our house, my father had two study rooms connected by a broad sliding door, which usually was open. One room he used for his scientific historical studies relating to classical philology, the other for his scientific interests, in particular mineralogy, botany and zoology. When our games with neighbours' kids in front of the house would become too noisy, he would knock at the window panes. This usually having only a brief effect, he soon knocked a second time, this time considerably louder. At the third knock, Helmut and I had to come to his room and sit still on a low wooden stool, back to back, up to one hour at 2 meter distance from his desk. While doing so we would see on a table in the other room the pretty yellowish wooden box that housed my father's big Zeiss microscope, which we were strictly forbidden to touch." (Ruska, 1970a, 1970b, 1987; Lambert & Mulvey, 1996).*

The account carries the suggestion of the forbidden fruit. As such there is little reason to believe that Ernst Ruska and his brother would not have grown up with an outspoken awareness of the significance of microscopes.

Nor let us forget Gabor's remarks already quoted, stating that "in 1927 nothing more than the words 'electron lens' were needed to start a real burst of creative activity", physicists asking themselves, "What can we do with electron lenses? Can we make electron microscopes?" On first sight, Gabor's account seems to be a bit exaggerated as so far, we have not even seen the words "electron lens" used by those involved at the Technische Hochschule, the place where Gabor himself came from in 1927. Nevertheless, even if Gabor had been carried away a bit, his perception of those days might indicate just as well that it was not that outrageous to start thinking of a practical application of electron optics—something like an electron microscope for instance. Altogether it appears to be justified, therefore, to designate the device with which Ruska made his first two-stage images as "first prototype of the electron microscope."

Since the subject of this part is the conception of the idea of an electron microscope, it would make sense to pause at this stage, and to continue the story of the developments at the Technische Hochschule in a second part, which would be dedicated to *materialising the idea.* After all, the birth of the very first electron microscope has just been painted. There is a clearer point, however, to leave off and that is the moment that the device received its actual name and the word *Elektronenmikroskop* started to appear in print, which happened within the following year.

1.5 Ruska's and Knoll's First Joint Article

After Ruska's first two-stage magnifications had made clear that tangible results could be obtained with electron optics, it became suddenly urgent to publish the results. As Lin has observed, there were no more entries in Ruska's lab journal after his historic experiment of 7 April 1931 until the completion of the first article by Ruska and Knoll. The first new entry in the lab journal was made on 27 April, one day before the submission date of the article (Lin, 1995). That joint paper was called *Die magnetische Sammelspule für schnelle Elektronenstrahlen* (The magnetic concentrating coil for fast electrons) and it finally appeared in the August issue of the *Zeitschrift für technische Physik* (Ruska & Knoll, 1931). The 'fastness' in the title was a reference to the high acceleration voltages that were applied. In the paper, the authors presented the results from the *Studienarbeit*—Ruska's first research assignment that he had finished two years before. As such, it was the very first article to provide an experimental confirmation of the theory of Hans Busch that electromagnetic coils act like lenses.[2]

From the article, it becomes very clear that the authors had begun to develop a true electron optical mindset by then, systematically approaching the electric coil as an optical device. They even realised that:

> *"The concentrating coil acts, in accordance with the Busch theory, as a 'convex lens for cathode rays' with the characteristic—which optical lenses do not have—of a freely adjustable focal distance."*

It is noteworthy that Ruska and Knoll already pointed this out when the first enlargements ever had hardly been made. Indeed, it is a quality that makes magnetic lenses essentially different from glass ones. In the case of a glass lens, its focal distance is always fixed. For each distance you need a different lens. That is why light microscopes have a revolving nose piece with different objective lenses, enabling you to choose between a number of fixed magnifications. In the case of a magnetic lens, however, you can change the electrical current in the coil instead. At the same time, the fact that the electron optical terminology was still very new and fresh, is revealed by the observation that quotation marks are used as soon as the concentrating coil is likened to a convex lens. Of course we do not know whether it was the editor or the authors who put them there.

Maybe the most important subject of the paper was treated in Section 5, where Knoll and Ruska discuss the benefits of an iron casing. It is interesting to see that their view on it had become far more articulate since the 1929 Studienarbeit. They now argued that the application of an iron casing would reduce the number of windings in the coil needed to obtain a field with equal strength, which would result in a substantial weight reduction of the coil. They deemed this to be relevant, as they say that the average concentrating coil weighed 3 kilograms approximately. There was also the brief mention of experimental findings this time: with a casing made of 2-mm-thick iron the concentric gap at the inside of the coil had to be 1 cm wide for the best result, which was a 33 percent reduction of the current needed—a result that can be translated to an equivalent reduction of the number of windings instead.[3] Subsequently they express

[2] The AEG Research Institute referred to the application of Busch's theory in 1928, but only used it to properly focus the electron beam without trying to validate the lens theory. See Rupp (1928).

[3] Note that the casing does not yet turn the coil into a polepiece lens, since the polepieces themselves are not yet present. These extrusions that further narrow the magnetic field were added to the concept a year later.

the expectation that thicker iron would reduce the optimal width of the gap and lead to a further reduction of necessary current. These additional experiments had not yet been performed, however, and this appears to indicate that these preliminary findings with an iron casing were rather recent. That would make sense, since Ruska did not obtain them during his Studienarbeit which ended in May 1929, while he had dedicated his second assignment to the electrostatic lens. It means that most likely he only started experimenting with casings somewhere in the first months of 1931, after obtaining his engineering degree. The fact that these recent results were now published together with the earlier findings from the Studienarbeit, might explain why Ruska started to believe in later years that he had already tested these casings in 1929. It should also be stressed that there is no mention yet of the possible reduction of the focal length as another possible advantage of an iron casing. The only additional advantage that they pay attention to is the elimination of the stray field which otherwise might interfere—as Gabor had already discovered—with the deflection plates if applied in an oscilloscope.

At the bottom of the journal's page on which this Section 5 on casings appeared, one finds a remarkable editorial note. Quite ironically, Section 5 had been left out in the original publication, and therefore the editor writes:

> *"This addendum appears in the September issue—although the subject does not belong to technical optics—because it is a part of the Ruska-Knoll contribution, which has accidentally been omitted from the August issue."*

With our present awareness that these casings are one of the most fundamental characteristics of electromagnetic electron lenses, it is even more ironic that the editor considered this subject to lie outside the field of optics. It illustrates that in spring 1931 electron optics was also a very new subject to the German *Zeitschrift für technische Physik*.

Again it should be stressed that there is no mention whatsoever of Louis de Broglie's wave–particle duality in the article by Ruska and Knoll. The whole idea of an electron lens is purely based on the work of Hans Busch. Finally, it is also worth noting that the authors were already referring to their next article in this first one, which carries the suggestion of a certain continuity between this first and the second. This is not a real surprise, as this next paper was going to be submitted on 10 September 1931, approximately four months after the first one. Before that, however, an important event took place, when the moment had arrived to make the experimental results officially public for the very first time.

1.6 The Cranz Colloquium

On 4 June 1931 Max Knoll gave a talk at the so-called Cranz Colloquium, named after Carl Cranz, a professor of ballistics who organized this well-attended series of public lectures at the Technische Hochschule (Ruska, 1979; Lin, 1995). In it Knoll presented the experimental confirmation of Busch's lens theory, together with the first two-step magnifications and some other optical observations which were going to be the subject of their following paper. In his description of these latest findings, again the display of electron optical thinking had become significantly more outspoken than it was in the article that Ruska and Knoll had submitted just five weeks before. Knoll explained for example:

> *"Furthermore, it should be possible to form images with two focusing coils for example, that is, with a lens combination. You see here the schematic arrangement of object plane, first focusing coil, first image plane, second focusing coil, second image plane. So, what we have here is, optically speaking, a Kepler telescope (...)" (Knoll, 1931; Ruska, 1979, 1984, 1986).*

This is pure optics, but in retrospect again, the comparison with a telescope sounds rather puzzling. He was describing the very first prototype of the electron microscope, as you might well say, and instead he called it a telescope. Neither was it the most likely comparison. Basically there exist only two geometric optical instruments, and these are the microscope and the telescope. However, you do not put metal grids in telescopes in order to magnify them, so for that reason alone it would have made more sense to compare the arrangement to a microscope. Knoll's choice of the telescope might reveal something completely different, though. Until then, both Ruska and Knoll had been very conservative in their terminology, probably wishing to avoid wild speculations, just as Hans Busch had refrained from discussing any sort of electron optical application in his papers. So, when finally the moment came for Knoll to make a comparison, it turns out that he chose the telescope, which must mean that the suggestion of an electron telescope was less outrageous than the suggestion of an electron microscope. As a matter of fact, as we will see later on, Knoll was not the only one who had been thinking in this direction.

1.7 Ruska's and Knoll's Second Article

Applied electron optics finally came to full blossom in the second joint paper of Ruska and Knoll, which was submitted on 10 September 1931 to the *Annalen der Physik* with Knoll as first author this time. It was a long article of 55 pages, called *Beitrag zur geometrischen Elektronenoptik*, which finally appeared in February 1932, divided between two successive issues of the journal (Knoll & Ruska, 1932b). From a historical point of view, it is without doubt their most important paper. In this paper they presented the drawing of Ruska's embryonic microscope on which Fig. 1 is based, and his very first two-stage image. These results were embedded in a thorough discussion of geometric electron optics in which they do not hold back anymore. The second of the six parts that made up the paper, was called "Imaging with diaphragms and electron lenses" for instance, while the last section of that part carried the title "Optical aberrations." It is also in this part that they present the two-stage magnification and other proof for the optical behaviour of cathode rays. They show for example that the size of a diaphragm will only affect the intensity of the image, but not other aspects such as its dimensions. To do this, they produced an image of the cathode surface. Markings on this surface proved that it was indeed a real image and not some sort of interference pattern. They also experimented with increasing currents in the electron lenses. For this purpose they made use of the same grids on top of the two lenses that had been imaged in the two-step magnification. An extra object was added half way between the two grids, which was a square diaphragm. According to theory, if you increase the current in a magnetic lens, you will reduce its focal distance. They tested this by using the bottom lens, which was positioned behind all three objects, as seen from the cathode. This way they were able, simply by increasing the current, to create step by step sharp images of the cathode, then the first grid, subsequently the square diaphragm and finally the second grid.

The article contains many more first observations. In Section 5 of the second part, the comparison with the microscope is finally made, which is the first time ever that the analogy had been submitted to print—and which is not the same, by the way, as the first time that it *appeared* in print, as will become clear in Section 3 of this study. In the same sentence, the lens immediately behind the first grid was likened to the objective lens of a light microscope, which is another 'first', at least in a scientific contribution and certainly in print. In the following section on optical aberrations at the end of the second part, "chromatic aberration" is mentioned for the first time ever. The authors even dedicate a complete subsection to it where they extensively discuss the differences in electron speed that are the cause of the chromatic error. In fact, they considered the chromatic aberration to be a bigger problem than spherical aberration.

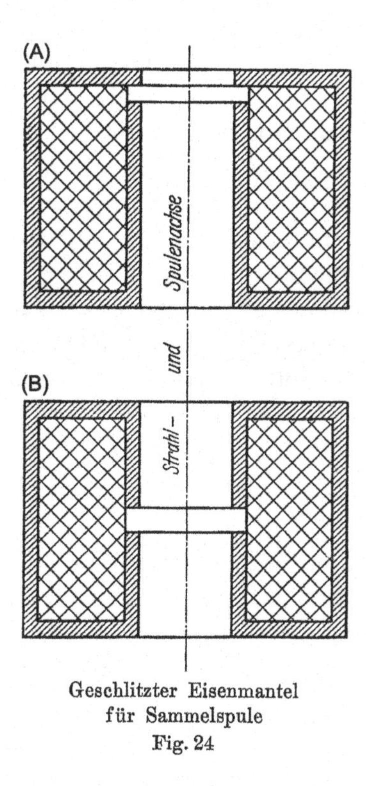

Figure 2 Encased electrical coils. The vertical line indicates the axis of the electron beam. The horizontal white bars indicate the slits in the two alternative encasements. From Knoll and Ruska (1932b, p. 648).

Undoubtedly the most historical 'first', however, comes in the third part, which deals with the possible "Application of geometric electron optics to electron tubes", as the title says. The second section of this part is called *Erzeugung grosser Elektronenbilder* (*Elektronenmikroskop*), meaning "Production of large electron images (electron microscope)". It is the first time ever that the word "Elektronenmikroskop" itself had been submitted to print, and here too we will see that this is not automatically the same as appearing in print for the first time. The most important topic in this section on the electron microscope is the need to reduce the focal length of the electron lens if you wish to produce large magnifications. This leads the authors immediately to the iron casing. As a matter of fact, they do not tell much more than they already had done in their previous article, but at least they finally present a drawing (Fig. 2). As you can see on the drawing, this is not a polepiece lens, since there are not yet polepieces that further narrow the passage at the position of the gap. The significance of a short focal length was subsequently illustrated with a theoretical example. They imagined a one-metre-tall tube with three lenses with a focal length of 2 cm and an image distance of 30 cm. Such a device could provide a tempting magnification of $(15 \times 15 \times 15) = 3375$ times. With a light microscope such large magnifications are out of the question, so it seems the example was meant to tease the reader's mind.

At the same time, they did not further speculate on any type of application apart from the possibility of forming enlarged images of the cathode itself, which they had already mentioned in the introduction of the section. The closest they got to what would become the future of the device was the remark that

> *"Very small original diameters (objects) are especially suited for magnifying in this fashion (...)."*

They were also aware, however, that high magnifications would require a very intense electron beam:

> *"Of course, large magnifications demand correspondingly high intensities of the imaging beam, if one is going to see anything at all on the electron sensitive coating."*

In practice, this has turned out to be a very relevant issue in later years. Enlargement is expressed on a one-dimensional scale, which only gives the increase of either width or height. In reality, however, if you enlarge something 1000 times, you actually magnify it 1,000,000 times, because the real magnification is two-dimensional. When making the width of an image 1000 times larger, the area increases $(1000)^2$ times. However, when magnifying an image a million times, the radiation that you use to project such an image, will be diluted a million times as well, because it needs to cover an area that is a million times bigger. If you want to magnify a 100,000 times, this diluting effect will even reach 10 billion times. In that case you will have to start with an immensely high intensity of electrons.

For the record, it should again be emphasised that there is no mention whatsoever of Louis de Broglie and his wave–particle equation. The whole concept of geometrical optics was still purely based on Busch's lens theory.

1.8 The Role of Max Knoll

Many historical reviews mention Knoll and Ruska in one breath as the inventors of the electron microscope, or at least as the two men who built the first one. This is very understandable as the earliest articles on electron optics, originating from Professor Matthias's High Tension Lab, were published by them together. Apart from that, Knoll had been Ruska's supervisor for a considerable time and there was this daily exchange of ideas in the Knoll Group, so it appears to be perfectly fair to credit both men equally for what had been achieved so far.

Notwithstanding this, there are a number of arguments to discriminate between their contributions. First of all, the actual results were all obtained by Ruska within the framework of his two assignments and successive preparations for his PhD. As Ruska told us in our 1985 interview:

> *"Knoll was sitting in the room in which the experiments were done. Evidently we were regularly discussing these matters. And because of that we also published together. But the experiments I did on my own. (...) He had more projects to look after" (van Gorkom & de Haas, 1985).*

For this presentation of their relation a confirmation can be found in the talk that Knoll gave at the Cranz Colloquium in June 1931. He told his audience for instance:

> *"Experiments in this direction were mainly done by E. Ruska, and have given experimental proof that the behaviour of electron rays (...) follows the laws of geometric optics very precisely" (Knoll, 1931).*

And a little later Knoll said:

> *"In the next image we see the experimental arrangement that Ruska used for testing the regularity of the image rendering."*

Otherwise, he systematically avoided using the personal pronoun "we."

But even more telling perhaps is the absence of any evidence that Knoll shared a belief in the possibilities of an electron microscope. The most striking instance, of course, was given by himself when he likened the first prototype of the electron microscope to a telescope. In the 1985 interview Ruska fully agreed that it was hard to understand why Knoll had said this, and in general Ruska did not think that Knoll had cherished high expectations of an electron microscope at the time, telling us for example:

> *"In these days the idea of an electron microscope that was better than a light microscope, was very vague. (...) I don't think that Herr Knoll really believed in a significant meaning of an electron microscope at the time, and quite soon afterwards he left, and went into the television industry. He certainly had no expectations of it at the time. (...) Neither did he ever claim any credit during later developments."*

Some significance should also be attributed to Knoll's failure to patent the electron microscope—an omission for which he could be held responsible since he was the senior. As we will see in Section 3, this error has had vast consequences. Mulvey suggested in a historical contribution from 1973 that this negligence had been due to Knoll's inexperience (Mulvey,

1973; Lambert & Mulvey, 1996) but in 1985 Ruska agreed with us that such an explanation does not hold:

> *"At the time, Knoll had definitely filed a patent already—as a matter of fact a patent on an electrostatic lens. That was already in 1929. So, indeed it was not as though Knoll did not know or understand anything at all about patents. But he had certainly not believed in the idea of a useful electron microscope with a substantially higher resolution. (...)*
>
> *Anyway, he did not have that much belief, or did not estimate the chances to be that great to put his own scientific work on the line. He was interested in other things."*

On 1 April 1932 Max Knoll left the Technische Hochschule and went to electronics company Telefunken where he was to supervise the development of television. In his position it was a logical step. Twentieth century television tubes stem directly from cathode-ray tubes and rely on the same type of technology as oscilloscopes to steer the electron beam from right to left and from top to bottom.

1.9 The Role of Ernst Ruska

The tentative conclusion that Knoll remained largely in the background, automatically implies that the earliest development of the electron microscope was mainly Ruska's work. Whether this is a correct representation of history is hard to prove or to disprove, although it appears to be plausible in the light of later developments. It is even more difficult to find evidence that it was Ruska who actually believed in a future electron microscope at the outset. In the light of this, it is interesting to take a closer look at a talk that was given by Ruska's father Julius, in an attempt to understand the intellectual background from which Ruska came. Ruska's father delivered the lecture on 5 June 1931, at the celebration of the 25th anniversary of the Berlin Society for the History of Sciences, Medicine and Technology—only a day after Knoll's first public announcement of Ruska's results at the Cranz Colloquium. As mentioned above, Julius Ruska was director of the Research Institute for the History of the Natural Sciences and was specialised in the history of the natural sciences in Islam. The memoir and bibliography published after his death in the journal *Osiris* show that his major subject of interest had always been the history of chemistry in Islam. However, the title of the lecture that he gave in June 1931 was "World View and Research Through the Ages" (Ruska, 1932). It was a remarkable demonstration of broad thinking on a scale that can hardly be surpassed.

In rough outlines Julius Ruska described two thousand years of scientific theory to reach the conclusion that our world view is just as infinite as the universe itself, whether it concerns infinity of largeness, infinity of smallness or even infinity of time. When we realise that father Julius was a specialist on Islamic culture and came from a classical background, we are struck by some very progressive statements, like:

> *"(...) It is especially typical of the new spirit of the times that advances also come from new apparatus, from instruments for observing and measuring, which make it possible to improve the observation of natural phenomena or to discover things completely unknown."*

This is followed a little later by:

> *"It may be that our astronomical world view has expanded into the infinitely large. But the broadening of our knowledge of nature into the infinitely small, which took place in the recent past, turns out to be even more amazing (...)"*

and by:

> *"In this way a physical–chemical world view has developed in the last decades, the great simplicity and profundity of which makes one pause, maybe with the following question on one's lips: what else is left to be discovered, what further secrets could be hidden behind these unbelievable insights."*

And subsequently he comes to speak of the microscope:

> *"... a completely new world, the world of the microscope, begins in the 17th century with Malpighi, Leeuwenhoek and Swammerdam (...) to occupy researchers and to astonish onlookers. The significance of the microscope for the expansion of our world view, for the knowledge of innumerable unknown, previously invisible forms of life, for the understanding of inner structures and developmental stages, but also for the understanding of inorganic nature, is a thousand times larger than the significance of the discoveries made with the telescope in the sky. No lecture can elaborate, nobody can sufficiently paint the extent of the revolution of our views on the living world that has been brought about by the microscope, this present of physics to biology, in the hands of researchers and doctors."*

Having a father like Julius Ruska, with such strong feelings about this 'present of physics to biology', must somehow have helped Ernst Ruska. The words of father Julius are especially remarkable since for years he had been very much opposed to the technical studies of his son Ernst. Julius was a true scholar and academic, and for that reason he considered a technical education inferior and certainly below family standards (Ruska, 1987; Lambert & Mulvey, 1996). This might also explain, incidentally, why in the end Ruska's parents supported Ernst's PhD, since that would at least bring him a respectable doctorate.

Of course, you can argue that Julius Ruska's interest in microscopy was already evident as he had this impressive Zeiss microscope standing in his study. But that does not automatically imply that the owner attributes a dominating historical and philosophical significance to it. Many scholars still consider instruments of secondary importance in relation to the development of theory. Neither is it self-evident that Julius Ruska should take the microscope as the finest example of scientific progress. There were many other developments to choose from. It is also noteworthy that he spoke about this subject as an expert on Islamic history, specialised in Arabic sciences. It would have been less of a surprise if he had dedicated his lecture to the oriental roots of the natural sciences. Therefore, Julius Ruska's appraisal of microscopy seems to come completely out of the blue. This also surprised Ernst Ruska, as he told us in 1985; he had only found out about this lecture a few years before we spoke to him. To Ruska, the discovery of his father's lecture was so significant that he took the trouble to get a copy of it and sent it to me by post some weeks after our interview.

2. ERNST BRÜCHE AND THE AEG RESEARCH INSTITUTE

In 1928—the same year that Matthias appointed Knoll as head of his new research team—the electronics company AEG founded a new research institute in the *Holländer Strasse* in Berlin-Reinickendorf, six kilometres north of the Technische Hochschule.[4] AEG stands for *Allgemeine Elektrizitätsgesellschaft* which is German for "General Electric Company." The fact that this is identical with the name of the well-known American company is no coincidence. In 1930, GE owned 25 percent of the AEG shares through its 100 percent daughter company International General Electric (Flaningam, 1945) In a personal document of 1945, the head of the institute's Physics Laboratory wrote about the *AEG-Forschungsinstitut*:

> *"The foundation had been encouraged by General Electric Company, which had a patent agreement with AEG and did not want to put in the needed developmental efforts alone anymore" (Lin, 1995).*

This head of the Physics Laboratory was the physicist Ernst Brüche, a pupil of the renowned Carl Ramsauer who had gained fame in 1920 with the Ramsauer effect, also known as Ramsauer–Townsend effect. This effect is regarded as the first experimental indication of the wave–particle nature of electrons. In later years Ramsauer would play a prominent role in German physics—from 1940 to 1945 he was President of the *Deutsche Physikalische Gesellschaft*. In 1928 he also happened to be the one who hired

[4] The institute's address *Holländer Strasse 31* is mentioned in a letter from Max Knoll to Ernst Brüche dated 8 October 1941.

Brüche for AEG, since he was appointed director of the new AEG Research Institute. Apart from that, Ramsauer had been an assistant of Philipp Lenard.[5] From this genealogy it easily follows why the Research Institute was not going to be a minor player in the new field of electron optics. As a matter of fact, Ernst Brüche would soon become Ruska's fiercest competitor for many years.

2.1 The Aeroplane Compass and the Aurora Borealis

As early as 1930, a fancy book appeared about the research and engineering at the new institute. It was called *Forschung und Technik*, published by Springer on behalf of the company and edited by Waldemar Petersen, a professor of electrical engineering and one of AEG's general directors (Petersen, 1930; Ramsauer, 1949). The book was certainly proof of AEG's keen eye for publicity with its impressive dimensions and 597 glossy images, which was a considerable number in those days, when printing was still a very elaborate process. The chapter in the book that covered the emerging field of electron optics was written by Ernst Brüche and called "Rays of slow electrons and their technical application" (Brüche, 1930). Slow electrons were generated by applying low voltages to cathode-ray tubes. In the case of AEG this meant voltages of less than 1 kV. In comparison, the lowest voltage used by Ruska at the High Tension Lab was 30 kV.

Brüche had very specific reasons to prefer slow electrons, as he explains in his article. In 1930 he envisaged the development of instruments, based on the cathode-ray tube, that would make use of the sensitivity of cathode rays to electrical and magnetic fields. Cathode rays are influenced by the earth's magnetic field for example, and also by all fields generated by electrical circuits, like the ordinary electrical wiring in your wall. This sensitivity of the cathode ray, however, depends on the velocity of the electrons. The slower they are, the longer the electrons experience the influence of the field and the more the cathode rays are deflected. The faster the electrons are, the less time remains for any interaction to take place. Brüche had also very practical reasons to use slow electrons as he considered high voltage tubes

[5] Ramsauer was appointed Honorarprofessor (unsalaried professor) at the Technische Hochschule Berlin in 1931. More on Ramsauer in: F. Fraunberger, "Ramsauer, Carl Wilhelm," *Complete Dictionary of Scientific Biography 15*. Detroit: Charles Scribner's Sons, 2008, pp. 490–491. *Gale Virtual Reference Library*, http://go.galegroup.com/ps/, accessed 27 April 2013. See also the chapter *Meine physikalischen Erinnerungen*, in Ramsauer (1949), pp. 99–130.

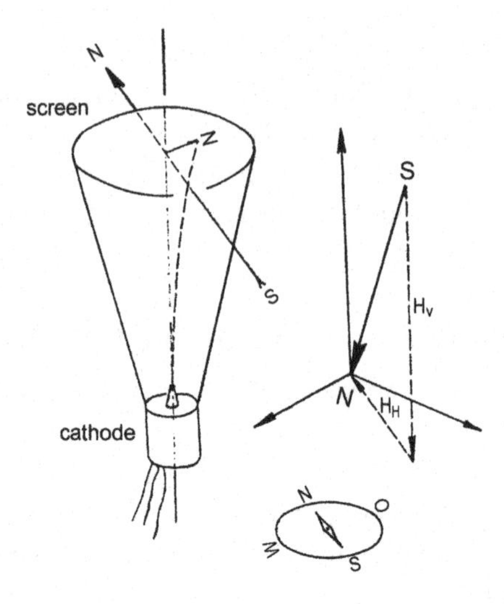

Figure 3 Schematic representation of a cathode aeroplane compass. 'Z' is the deflection. From Brüche (1930, p. 37).

> *"dangerous, expensive and difficult to promote with all their auxiliary devices" (Brüche, 1930).*

From the commercial perspective that you should keep a technical design as simple as possible, this point of view of course made perfect sense.

One of the instruments that he envisioned was an aeroplane compass, which was clearly his favourite as he dedicated more than five large pages of his article to its description and the principles on which it was based. Actually, the compass was little more than a cathode-ray tube placed in a vertical position, with the viewing screen facing upwards and the electron gun touching the floor of the aeroplane. The rest was automatically done by Mother Earth herself. The earth's magnetic field has a certain shape which can be visualised as field lines. These lines come out of the earth at an angle that depends on where you are. Free electrons will follow the direction of the magnetic force lines and therefore the beam in the vertical cathode-ray tube will not hit the screen right in the centre, but will be bent sideways, as shown in Fig. 3. In theory, the direction of this deflection can be used for orientation. In practice, proper interpretation was rather complicated. Brüche also presented less outrageous designs. The more practical applications were a detector for electromagnetic fields and devices to measure currents and voltages.

Where the simplicity of design was concerned, Brüche was very consistent in applying this approach to the creation of cathode rays as well. In this case, it was his philosophy that the best lens would be no lens, and therefore he was delighted by what he called *Fadenstrahlen* (thread beams) (Brüche, 1930). If enough residual gas is present in a cathode-ray tube, the electrons in the cathode rays will ionise part of the gas atoms. With a carefully chosen amount of gas, a trail of big and slow positive particles can be created that will hold the beam of negative electrons together. As an additional attractive feature, this long trail will light up as well, since some of the gas atoms become exited, i.e. energized, instead of ionized, after which they emit their surplus energy as light. With such beams you do not need any concentrating devices, since the beam does not diverge as it would normally do. Brüche states in his article that they were able to create thread beams more than 1.5 m long. They had a diameter of just 1 mm and were created with a tension of only 200 volts. His description of the beams is illustrated with no less than 12 photos.

Apart from the practical use of such self-concentrating beams, the research of these beams could also have a more fundamental relevance, Brüche believed, referring twice to the northern lights theory of the Norwegian scientist Carl Störmer. Störmer's work was based on Kristian Birkeland's assumption that the northern and southern aurorae are caused by the emission of "cathode rays" by the sun (Störmer, 1929). Störmer had widened this view to the emission by the sun of other charged particles such as the alpha particles. An alpha particle is a helium atom stripped of its two electrons and therefore positively charged. In 1929—so only a year before Brüche's contribution—Störmer had published a report on late echoes of short wave radio signals, which was meant to prove the validity of his theory. These echoes would follow primary radio signals with delays varying from 4 to 30 seconds—a phenomenon that could only be understood by assuming that the echoes travelled distances of millions of kilometres. Störmer's streams of charged particles from the sun are nowadays known as solar winds and will spiral along the field lines of the earth's magnetic field, creating a kind of electrical shield. At the earth's poles, where the magnetic field lines penetrate the atmosphere, the shield will show up as the aurora. Under certain circumstances, however, this electrical shield can also behave as a kind of mirror that reflects the short wave radio signals. Störmer estimated that the radius of the magnetic field would be over two million kilometres for electrons, which sufficiently explained the recorded delays of the radio echoes, he thought.[6]

[6] In fact, Störmer described the Van Allen Belts, as they are known nowadays. Current insights indicate that their radii are much smaller than Störmer's estimation.

One can imagine that these kinds of studies triggered the imagination of a physicist like Brüche. After all, it is the only example on our planet of a natural phenomenon that can be explained by cathode rays. Lin (1995) mentions that Brüche even made a trip to the Auroral Observatory in Tromsø, Norway, in spring 1931 and published a film about the aurora borealis later that year. As a matter of fact, Birkeland and Brüche had not been the first to make the connection between the aurora and cathode ray tube discharges. This had, quite remarkably, already been done by William Watson, the very first builder of a discharge tube. In his "Account of the phaenomena of electricity in vacuo," published in 1752, he observed

> *"(. . .) here the coruscations were of the whole length of the tube between the plates; that is to say, thirty-two inches, and of a bright silver hue. These did not immediately diverge as in the open air, but frequently, from a base apparently flat, divided themselves into less and less ramifications, and resembled very much the most lively coruscations of the aurora borealis" (Watson, 1752).*

2.2 Electron Optics

By now, you will probably be wondering what all of this has to do with geometric electron optics—especially when taking into account that Brüche was going to challenge Ruska relentlessly within less than one and a half years. The electron optical relevance of Brüche's article from 1930 is found in his discussion of ways to focus a cathode ray. As we already saw, he preferred the self-focusing property of the thread beam, but in the preamble to his description of these beams he also paid attention to the magnetic and electrical lens in relation to Busch's lens theory. As such, he spoke of "convex electron lens" and of "imaging" with such a lens as early as 1930, which makes it the first time ever that the words "electron lens" appeared in print (Brüche, 1930). The words were put between quotation marks by Brüche to underline their hypothetical nature, as Ruska had done when he discussed the "focal distance" of a concentrating coil for the first time in 1929. The imaging was Brüche's reference to the possibility of producing images of the cathode itself. Brüche also deserves the credit of having merged Busch's phrases *geometrische Optik*, *Elektronenbahnen* (electron trajectories) and *Optik der Kathodenstrahlen* (cathode-ray optics) into *geometrische Elektronenoptik* (geometric electron optics) for the very first time, which he does in a footnote. These semantic exercises are especially noteworthy, since Busch's theory had not yet been officially confirmed experimentally—the paper of Ruska and Knoll that would do so was to appear only in August 1931.

In this regard it is certainly ironical that the new electron optics does not appear to have been Brüche's primary interest. Indeed, his whole contribution contains many indications that he had only partially grasped the new electron optics at that moment. He mentions for instance that a magnetic field causes refraction, comparable to the action of a glass lens, while an electrostatic field would only give a reflection in the way a mirror does. He refers to the wave–particle nature of electrons, but at the same time he says:

> *"In contrast to rays of light, however, rays of electrons are* material carriers *of electrical charge" [emphasis by JvG].*

And when discussing the relation between the accelerating voltage and electron speed, at the one hand he presents a graph that flattens at the top to show that the speed of light is the ultimate limit. Although this limit is an immediate outcome of Einstein's relativity theory, he also gives equations from classical physics, which ignore that very same relativity theory and will therefore never be able to produce the mentioned graph.

The references to the wave–particle nature of electrons are especially intriguing. We have seen that Ruska and Knoll made no reference to de Broglie's wave–particle equation. Brüche too does not elaborate on the equation, nor does he mention de Broglie, which is certainly surprising since he is not an electrotechnical engineer, but a physicist specialised in electrons—and even a student of Carl Ramsauer, the man who was one of the first to discover that electrons do not always behave the way you would expect solid particles to do. In this respect it is even stranger to know that in 1928 the AEG Research Institute had already published an article by Emil Rupp to prove the validity of de Broglie's thesis for slow electrons. His results of experiments with electron diffraction showed a 98 percent agreement with the values predicted by de Broglie's equation (Rupp, 1928). In February 1931, Rupp even submitted a second article in which he had repeated his diffraction experiments with fast electrons (Rupp, 1931). In this second paper, he claims to have obtained a 99 percent experimental agreement with de Broglie. Since Brüche was head of the Physics Laboratory, you would expect him to be familiar with these two papers. One has to conclude the de Broglie equation was too esoteric to him at the time—and, so it appears, not only to him.

2.3 Brüche's Electron Microscope

Notwithstanding Brüche's strong emphasis on instruments to detect electromagnetic and electrostatic fields, research at the AEG Physics Lab did start to move in the direction of electron optics within the following year. This development was made public by Brüche in the issue of 15 January 1932 of the highly esteemed German journal *Die Naturwissenschaften* (Brüche, 1932a). His preliminary report of only half a page was published in the section "Short Preliminary Notices", which was a quick way of securing priorities. It was dated "November 1931", and carried the pregnant title *Elektronenmikroskop*. This title alone laid the foundation for a long lasting conflict that was soon going to develop between Ruska and Knoll on one side and Brüche on the other. The essence of the conflict was as simple as it was irreconcilable: on 10 September 1931 Knoll and Ruska had been the first to submit the words "electron microscope" to print, but the paper only appeared in February 1932, while Brüche had been the first to have the words actually appear in print a few weeks earlier, on 15 January 1932, although he had submitted his announcement in November 1931. Both parties therefore claimed to have been the first to coin the name for the new electron optical device. Fifty-four years later, in 1985, Ruska was still annoyed about this affair, as he then told us:

> *"Herr Knoll had told Herr Brüche on 13 November '31 in which fields we were active, and that an article had been submitted by us. It was this article about electron optics (Knoll & Ruska, 1932b). Herr Brüche, who had just obtained the first images of a cathode, then published a short notice in* Die Naturwissenschaften *and this notice appeared in print sooner than our sixty-page-long [sic] article in* Annalen der Physik. *And from then on he has always stressed that he was the first to mention the electron microscope in public" (van Gorkom & de Haas, 1985).*

As a matter of fact, Brüche had been fair enough to acknowledge Knoll and Ruska in his announcement. His contribution starts thus:

> *"From Herr Knoll I learned that in the High Tension Laboratory of the Technische Hochschule experiments have been carried out to take enlarged images of cold cathodes, diaphragms and grids. A publication about this by Ruska and Knoll ('Contribution to geometric electron optics') is in press at* Annalen der Physik. *As the AEG Research Institute has been working in the same field too for about a year, and since useful results have been obtained here as well, a notice appears to be appropriate, especially as an early conclusion of the whole project is not to be expected, considering the broad definition of our research goals.*

One of these research goals was:

> *"to build an 'electron microscope' with very strong magnification to follow the emission process of an oxide cathode."*

The announcement was illustrated with a 60-fold and an 80-fold magnification of the surface of a cathode. The emission process at the surface of cathodes was known to be very irregular and would change over time, so as such this research goal was very legitimate. Brüche did not provide any further details about what his "electron microscope" looked like or how it worked, except for stressing the fact that slow electrons of several hundreds of volts had been used in contrast to the high tension experiments by Ruska and Knoll. This explicit statement is in line with the position that Brüche already took on slow electrons in the AEG book of 1930. At the same time, however, it is something of a contradiction to prefer to work with slow electrons, while it is one of the major goals to build an electron microscope with *very strong* magnification.

At this very early point in the history of the electron microscope it is hard to determine where this sudden move of Brüche came from. The fact that his announcement of an electron microscope is very short and its statement that the publication of experimental findings is not yet to be foreseen, conveys the impression that experiments had hardly started. In theory it is even possible that the published magnifications of the cathode surface had only been produced after Brüche had spoken with Knoll on that 13th of November. After all, his announcement is dated "November 1931", so it does not reveal the exact submission day. Only nine years later, by the end of 1940, did Brüche provide a few more details about this episode in a historical contribution, which was published in the company journal *AEG Mitteilungen* (Brüche, 1940). Here he wrote that his co-worker Helmut Johannson had been the first at AEG to realise that the images which appeared under certain circumstances on the fluorescent screen of a cathode-ray tube were images of the cathode. The photograph that Johannson produced on that occasion is supposedly the first electron image ever of a cathode surface. Once more, however, Brüche fails to provide an exact date, which makes it difficult to verify this story. All we know for sure at this stage is that there was another person who claimed to have coined the word "Elektronenmikroskop" in 1931, which is the right moment to move on to the next rival.

3. REINHOLD RÜDENBERG AND SIEMENS

In the same section of *Die Naturwissenschaften* in which Brüche's announcement had been published in January 1932, another announcement appeared that carried exactly the same title *Elektronenmikroskop*. It appeared nearly half a year later, in the issue of 8 July, and had been dated 7 June 1932 by its author Reinhold Rüdenberg (Rüdenberg, 1932). Since 1923, Rüdenberg had been the Chief Electrical Engineer of Siemens-

Schuckertwerke in Berlin, a division of Siemens, which was located in an area of Berlin which is still known as Siemensstadt, some five kilometres to the north west of the High Tension Lab, where Knoll and Ruska worked. The Siemens company as a whole was a major competitor of AEG, since it too was a large manufacturer of electrical products. Rüdenberg was "in charge of further developments in the field of electric high-power engineering" as he writes himself in a memoir from 1959 (Rüdenberg, 2010). In 1927, he had also been appointed *Honorarprofessor* (unsalaried professor) at the Technische Hochschule (Rüdenberg, 2010; Ruska, 1986),[7] so practically speaking he was an immediate colleague of Ernst Ruska's professor Adolf Matthias as well (von Weiher, 2008).

The announcement that Rüdenberg published in *Die Naturwissenschaften* in the summer of 1932 was only one paragraph long:

> *"As several parties have published proposals for the construction of electron microscopes lately, I venture to point out that within the Siemens Group too work has been carried out in this direction for some time, on using magnetic or electric fields with electron beams or proton beams as microscopes or telescopes. As goal we especially envisage to image enlargements of resting or moving submicroscopical objects (...). Although our fundamental patents reach back to May 1931, detailed publications will not be submitted until concrete development has progressed." (Rüdenberg, 1932).*

This notice on behalf of Siemens is essentially a priority claim, especially challenging the claim that Brüche made on behalf of AEG six months before, and in that respect the two claims show a certain similarity. Both Rüdenberg and Brüche wish to make it very clear that their companies are working on something called an "electron microscope" and both of them have to take refuge in the announcement of possible future publications as they were not in a position yet to publish full results—for whatever reason. As a matter of fact, "detailed publications" were never to appear in Rüdenberg's case. His "fundamental patents", however, were certainly there.

3.1 The Early Siemens Patents

The history of the early Siemens patents on the electron microscope[8] is a complicated one. It has been described in varying degrees of detail by several authors, among whom are Thomas Mulvey (1973), Ernst Ruska (1984,

[7] See also: Editorial, Weltkraftkonferenz, *Die Umschau* 34, issue 24 (1930) 486.

[8] I call these patents "the early Siemens patents" to discriminate them from the Siemens patents on the electron microscope that were filed from 1938 onwards.

1986), Martin Freundlich (1994), Qing Lin (1995), Reinhold Rüdenberg himself (2010) and Rüdenberg's son Gunther together with Reinhold's grandson Paul Rudenberg (2010). For that reason this discussion will be primarily restricted to those topics that shed most light on the priority issue.

In total there are eight German patents that were granted to Siemens, have to do with electron microscopes somehow and mention Rüdenberg as the inventor (Appendix A, Group I). This information immediately reveals that in all cases the patents describe so-called shop inventions, meaning that the inventions had been part of the job of the employee who is named as the inventor. As a result the company is the owner of the patents, since it has already paid its employee for the efforts that he made, and self-evidently it paid for the patent applications as well. Apart from the eight patents in Germany, there are eight similar ones which were granted in Austria, France, Great Britain, Netherlands, Switzerland and USA (Appendix A, Group II). Furthermore there exist three related patents which mention co-workers of Rüdenberg as inventor (Appendix A, Group III) and finally there are another three which mention no inventor at all (Appendix A, Group IV). This brings the total of Siemens patents from 1931, 1932 and 1933 that have something to do with electron microscopes to 22, of which no less than 16 name Rüdenberg. One reason why the history of these patents is complicated—if not confusing—is that there is no one-to-one relation between the most important patents and their original applications. Even Rüdenberg himself did not make an attempt to solve that puzzle. In the memoir of 1959, he discussed the two patents that were granted in the USA extensively, but when it came to the original ones, he restricted himself to the single statement that

> *"The German patents themselves have been greatly retarded, due in part to Nazi measures, to the Second World War, and to its aftermath, and were finally published in unduly subdivided shape from 1950 to 1954" (Rüdenberg, 2010).*

The reference to the Nazis is a recurrent theme in Rüdenberg's story, since he and his family were Jewish and had therefore been forced to leave the country in 1936. His son Gunther and grandson Paul tried to elaborate in more detail on the relation between the published patents and corresponding applications in their 80 pages long commentary on their father's and grandfather's memoir. They too, however, write:

> *"In Germany, the final fate of the [first] two applications was a prolonged and complicated one. (...) In the years 1936 to 1939, as the result of interactions between*

> *the Reich patent office and Siemens after Rüdenberg had left Germany, these applications were reworked and divided into five parts, with some important changes in language.*
> *The original applications of May 30, 1931, have, regrettably, not been located either at the Siemens Archives or the German patent office" (Rudenberg & Rudenberg, 2010).*

For this reason, I will take the actual patents themselves as point of departure. A second argument to do so is the fact that they are also the documents that matter most from a retrospective point of view, as it is their legal status that has played the largest role in discussions about priorities.

The eight German Siemens–Rüdenberg patents were applied for in a period spanning approximately 15 months. Three carry the priority date 31 May 1931 (Nos. 895 635, 889 660 and 906 737). One has the priority date 27 June 1931 (No. 916 838), another one 28 June 1931 (No. 911 996), two are dated 31 March 1932 (Nos. 916 839 and 916 841) and the last one carries 13 August 1932 (No. 915 253) as priority date.[9] So, except for the last patent, all have a date prior to Rüdenberg's little announcement in *Die Naturwissenschaften*. Notwithstanding the lack of clarity concerning the actual applications for these eight patents, the Rudenbergs were able to reproduce in their comment of 2010 an old undated list from Siemens, which supplies several original application dates. According to this list, two applications had been made on 30 May 1931, one on 26 June and one on 27 June 1931 and two on 30 March 1932. These dates happen to match with the above-mentioned priority dates insofar that there is a difference of exactly one day in each case. The dates are also in perfect accordance with the priority dates of all eight patents outside Germany. The only application that is clearly missing from the list is the one that would match the eighth patent of 13 August 1932, as the authors already point out themselves. That missing one would otherwise have brought the total of applications to seven. When considering the quote above that says that the first two applications resulted in five patents, one would even expect ten German patents instead of eight. As a matter of fact, this number of ten can be explained, since father, son and grandson Rüdenberg also took into consideration two of the Siemens patents that do not supply an inventor's name and which were never officially published, although they carry a patent number and are available through the German patent office.

[9] The two patents of 31 March 1932 are supplements to the 27 June 1931 patent, as is mentioned in the headings of the two patents.

By now it will have become clear that this whole history is complex indeed. The most important conclusion at this stage, however, is that apparently two applications were filed with the German patent office on 30 May 1931, which were split in five separate patents. Three of these patents carry the name of Rüdenberg as inventor. After 30 May at least four more applications were made, which resulted in five other Siemens patents that mention Rüdenberg.

That date of 30 May 1931 is crucial, since it was five days before Max Knoll gave his lecture at the Cranz Colloquium in which he presented for the very first time to the outside world the results that he and Ruska had obtained at the Technische Hochschule. For this reason, the priority set by the first two patent applications is seen as decisive for discussions on the priority of invention of the electron microscope. But before we move on to that discussion, we will have to take a closer look at the nature of the actual claims. In this respect it is already illuminating to examine the titles of the patents, especially since these titles are summaries of the primary claims made by each of them. If we limit ourselves to the three German Siemens–Rüdenberg patents with priority date 31 May 1931, we find the titles "Arrangement to influence the path of electron rays by means of electrically charged field diaphragms" (No. 889 660), "Arrangement for the magnified imaging of objects by means of electron rays and by means of electrostatic or electromagnetic fields that influence the path of electron rays" (No. 895 635) and "Arrangement for the magnified imaging of objects by means of electron rays" (No. 906 737).

The first patent, number 889 660, is basically a description of an electrostatic lens and as such less interesting since Max Knoll had already been the first to patent a similar concept in 1929. In Rüdenberg's case, however, it is the explicit purpose of the lens to use it in a microscope, as the patent literally says:

> *"In particular it is possible to use it for constructing a microscope that incorporates direct or reflected electron rays, and that allows for much stronger magnifications than our optical microscopes with their limits set by the wavelength of light."*

This is undeniably the concept of an electron microscope, although that specific name is not given to it yet.

Patent 906 737—"Arrangement for the magnified imaging of objects by means of electron rays"—seems to have been primarily intended to patent the concept of making enlargements. In reality, it pays substantial attention to an electrostatic lens which consists of a single diaphragm—in other words, a design that is nothing more than a charged sheet of metal with a hole in it. As Lin already pointed out in 1995, such a construction cannot be used as a lens for fundamental physical reasons. To make a diaphragm work like a lens, you will need to sandwich it between two additional diaphragms that carry a different charge. As a matter of fact, such a triple diaphragm is described by Rüdenberg himself in the patent 889 660, which has just been discussed. Apart from presenting a non-working lens in patent 906 737, Rüdenberg also attributes to it the ability to be used as either a convex or a concave lens, depending on the sign of its electrical charge. Even with a triple diaphragm this is not possible, so altogether this patent gives the strong impression that Rüdenberg patented ideas that had not yet been tested in the laboratory. Notwithstanding this, the text of the patent repeats in connection with the electric lenses that

> *"Magnifying glasses, microscopes and telescopes constructed in this way allow for a magnification range which is substantially wider than it is for optical instruments, of which the resolution is limited by the wavelength of light. This limitation does not exist for magnifying glasses that work with electron rays."*

It is interesting to see that there is no reference whatsoever to de Broglie's particle–wave equation and the limits that do follow from it. In this respect it is the same pattern as we already saw with the Knoll Group and Brüche.

It is also worth noting that a substantial part of patent 906 737 deals with the issue of creating beams of electrons with sufficiently uniform speeds, as Rüdenberg was aware that differences in speed within the same beam will cause chromatic aberration, although he does not use this term yet. That was going to be done by Knoll and Ruska for the first time, four months later (Knoll & Ruska, 1932b). Finally, it deserves to be acknowledged that the patent not only refers to the imaging of cathodes themselves, but also to the imaging of objects that either reflect electrons or are capable of letting electrons pass through them.

Patent 895 635—"Arrangement for the magnified imaging of objects by means of electron rays and by means of electrostatic or electromagnetic fields that influence the path of electron rays"—is the third Siemens patent with priority date 31 May 1931 that mentions Rüdenberg as inventor. The title is essentially a combination of the two other patent titles, so apparently the patent was meant to make it unambiguously clear that the idea of producing magnifications with electron rays, and the idea of influencing the path of electron rays by means of electromagnetic or electrostatic fields belong together. In fact, this combination of wanting to magnify with electron rays and using electron lenses to do this, more or less defines the electron microscope. Nevertheless, the text of the patent itself does not contain anything that is really new in comparison to the other two. The impossible concave lens is mentioned again, just like the suggestion to use such lenses to build a microscope or telescope and the belief that there would be no limit to the resolution if you magnify by means of electron rays. A little new detail is the likening of the lens behind the object to an "objective lens". This term was independently introduced by Knoll and Ruska in their large article on geometric electron optics which they submitted on 10 September 1931, a little more than three months after the submission of patent 895 635. Attention is also drawn by Rüdenberg's description of the electron image as a shadow image. This strengthens the impression that Rüdenberg was not aware of de Broglie's thesis. With the knowledge of the particle–wave duality lacking, electrons had to be particles and particles are expected to produce shadow images, just as the tiny droplets of spray paint will produce a shadow of an object and not an optical image. It should be added, though, that we cannot know for certain that the word "shadow" was also used in the original submissions, since we do not have access to them. The same can be said for the word "objective".

It is not my intention to perform a detailed examination of the two other Siemens patents that carry 31 May 1931 as priority date (patents 754 259 and 758 391) since they do not mention an inventor and actually do not offer any remarkable new details. However, a few observations can be made that have relevance to this discourse. First of all, apparently the two patents are considered by Rüdenberg to be derived from his first two patent applications, since they contain the same images and therefore the same explanations of these images in the text. A second remark is that the patents are only available from the German patent office in the form in which they were originally submitted. The submitted texts do not reveal a substantial difference in language or phrasing to the final form in which the three other patent texts were published. In the third place, it is noteworthy that both these patents make no mention whatsoever of the possibility to magnify. Patent 754 259 describes the electric lens as a focusing device and patent 758 391 describes the lens as merely an imaging device. As a result, the basic concept of using electron beams to magnify images becomes even more strongly linked to the name of Rüdenberg, as far as the Siemens patents are concerned.

Because of that, some attention should also be paid to the three patents that name someone else as inventor. German patent 659 359 gives the name Ernst Lübcke and is supposed to belong to the other patents since it appears on the Siemens patent list that is reproduced in the comment by the Rüdenberg family. The patent describes the idea of creating an electron arc and using this arc as a cathode. It seems therefore to be only faintly related to the electron microscope. A second Lübcke patent is also mentioned on the list. It is described as having to do with the use of reflected electron beams, but it is not possible, however, to find a patent in the database of the German patent office that even remotely matches this description.

The two other patents that were granted to Siemens and mention someone other than Rüdenberg are attributed to Richard Swinne and have a far clearer connection with the rest. Patent 915 843 is titled "Arrangement for imaging objects" and claims that other sorts of charged particles can be used as well. This claim is completely based on de Broglie's equation and illustrated with several calculations of wavelengths. The priority date is 18 June 1932, which makes the patent rather worthless since, a year before, the idea was already mentioned in a patent by Léo Szilárd as we will see in the next section. Patent 916 840 carries the same priority date and is titled "Arrangement for the magnified imaging of objects in two stages," which is not what you would expect. In this patent, stage one of the magnification is done with electron rays and stage two is a light optical magnification of the screen image. As a matter of fact, the patent is presented as a supplement to the Rüdenberg patent 916 839 in which the concept had been described for the first time. This patent was rather futile as well, since this idea had also been suggested by Szilárd in his patent a year before.

As far as the other 14 patents are concerned, they do not really add anything new to this discussion. If we simply take a look—just for the sake of history—at the patent titles we notice that most of them are variations on the two major themes 'arrangement for the magnified imaging of objects' and 'arrangement to influence the path of electron rays.' This observation is at least in good agreement with the assertion of the Rüdenberg family that originally it all started with two applications.

3.2 Rüdenberg's Source of Inspiration

It is now urgent to enquire where Rüdenberg's inspiration is believed to have come from when he decided to make his first two applications. As we saw, he wrote in *Die Naturwissenschaften* with regard to electron microscopes: "I venture to point out that within the Siemens Group too work has been carried out in this direction for some time". We also saw, however, that it should be seriously doubted that the patented electron lenses were ever subjected to laboratory tests. As a matter of fact, the Rüdenbergs themselves have provided us with a range of rather ambivalent messages in this respect.

Reinhold Rüdenberg and his family ended up in the USA in 1938 and in 1943 he wrote a letter to the editor of the *Journal of Applied Physics* which was published with the ambitious title "The Early History of the Electron Microscope" (Rüdenberg, 1943). In this letter he claims publicly for the first time to be the inventor of the electron microscope. When it comes to the actual work that was done, he writes:

> *"Since already the first investigations turned out to be very promising, the principal results were expressed in patent applications assigned to the German industrial firm with which I was formerly connected."*

So, here again the impression is raised that at least some experiments had been performed before the first patents were filed.

In the following years, Rüdenberg started to write and rewrite his memoirs of which the final draft of 1959 was only published in 2010, 51 years after his death. He paints quite extensively how he slowly came to realise the concept of an electron microscope. He presents the logic of the lines of reasoning that he had followed as a succession of steps which he had made, while it remains completely unclear whether these steps were based on any experimental outcome, or had been mental projections instead. He concludes this history with the sentence:

> *"All this was found out, when I brooded over viruses in the winter of 1930 to 1931."*

In the same period he made a successful attempt to have the ownership of the two US Siemens–Rüdenberg patents transferred in his name, based on the argument that the invention had actually been completely his own and had nothing to do with Siemens, but that he had been forced by the circumstances to assign his rights to the company.[10] On that occasion, in 1947, Judge Charles Wyzanski of the District Court in Boston, Massachusetts, asked Rüdenberg whether he had incorporated his theory in any way in any kind of apparatus to demonstrate it. Rüdenberg's answer was a straight "No", and he subsequently explained:

> *"I was very used to think an invention through to the very end before it was put into the real material by building an apparatus (Rudenberg & Rudenberg, 2010).*

[10] Rudenberg vs. Clark, 72 *Federal Supplement* 381 (1947).

This conversation is quoted in the comment that the Rudenberg family wrote and is introduced by themselves with the words:

> *"Rüdenberg's microscope design was based on ideas he developed over the winter of 1930–31. In the pattern of other Rüdenberg inventions, he had not yet built the apparatus at the time of his patent applications."*

But the ambivalence about the existence of any experimental work is so great that even in this same family comment one can read six pages earlier, in reference to the patented electric lenses:

> *"This has remained one controversial aspect of his design, suggesting to some that his ideas were patented before enough [sic] experiments were performed."*

From all this, we may safely conclude that Rüdenberg's patents were *not* based on any research that was going on at Siemens. So, where did he find his inspiration then? In the letter that he wrote to the *Journal of Applied Physics* in 1943 he also mentioned for the first time that his inspiration had been "a case of virus disease in my family" which had prompted him to find ways to learn more about viruses. Since he had understood that viruses are too small to observe with a light microscope, he concluded that a better microscope was needed. In the 1947 court case it became clear that the actual disease was infantile paralysis which had affected Rüdenberg's youngest son. In the 2010 comment by the Rudenberg family many more details are given, including fragments from an unpublished memoir by Rüdenberg's wife Lily Minkowski, who was a biologist. According to this account, Gunther's brother Hermann is believed to have contracted polio during a family trip to the Dutch coastal province of Zeeland in the summer of 1930. Altogether, the details provided make it certainly plausible that Rüdenberg was very worried about his son in the autumn of 1930.

It is a long way, however, from experiencing the acute need to invent a better microscope to actually coming up with a cathode ray tube with electron lenses to accomplish that task. Rüdenberg himself, as well as his sons, have gone to great lengths to make clear that he was rather brilliant and also possessed the necessary expertise. He had, for example already completed his doctoral thesis by the time that he finished his bachelor's, he had studied alongside Hans Busch and he had been taught by Emil Wiechert, the man who had introduced the electromagnetic concentrating coil. In his professional life, Rüdenberg enjoyed much esteem. Gabor (1974) introduces him as "Reinhold Rüdenberg, the very distinguished Chief Electrical Engineer of Siemens-Schuckertwerke". Mulvey declared in 1973:

> *"Rüdenberg had a reputation (...) for going straight to the heart of the problem. He had an extremely wide grasp of contemporary science (...)."*

And Max Steenbeck, a former personal assistant of Rüdenberg, wrote in 1977:

"Rüdenberg, who was more a theoretician than an experimental physicist, possessed an extraordinarily quick, determined, and rational way of thinking that was based on comprehensive, systematic, and always available knowledge regarding the conditions and limitations of the high-voltage field, but his knowledge was not at all restricted to this field. (...) He also possessed the ability—which he consciously exercised—of generalizing information pertaining to other areas of research, and from the information derived, he deduced possibilities for new developments in his own field." [11]

This solid reputation makes it certainly more likely that Rüdenberg was able to come up with the idea of an electron microscope. On the other hand, he too had no apparent knowledge of de Broglie, so it cannot have been that self-evident to him that rays of electrons would exhibit optical properties. The impressive size of the mental leap that Rüdenberg made, was accidentally made clear by Judge Wyzanski in his charitable verdict of 1947, when summarising the invention's history:

"What happened in 1930 and 1931 can now be told. In the fall of 1930 Professor Rudenberg's younger son was stricken with infantile paralysis. The father, deeply concerned about that illness, learned that doctors knew that it was carried by a virus of poliomyelitis, but a virus so small that it could not be studied under the lens of any existing microscope or any microscope which would depend for its operation upon light waves. The trough between light waves was so much larger than the virus that observations were impossible. Thereupon Professor Rudenberg set himself to devise a new type of microscope. First he thought of an X-ray microscope. This proved impractical. And finally he hit upon a microscope based upon electronic principles. (...) In accordance with what was his usual practice, Professor Rudenberg merely drew certain sketches of this invention. He did not embody it in any material apparatus." [12]

So, first Rüdenberg had to hear that it was a virus, than to learn that this virus is too small for a light microscope, than to discover that X-rays would not do the job and finally to come with the alternative of fitting a cathode-ray tube with electron lenses, and this all in approximately six months time without any experimental confirmation. I assume that most scientists and science historians will have some difficulty with the credibility of this story, although it would make a good film script.

[11] Max Steenbeck, cited in: Ernst Ruska, The Emergence of the Electron Microscope (Connection between realization and first patent application, Documents of an invention), *Journal of Ultrastructure and Molecular Structure Research* 95 (1986) 3–28, p. 24.

[12] Rudenberg vs. Clark, 72 *Federal Supplement* 381 (1947), p. 384.

3.3 Doubts

Two people who certainly did not believe that Rüdenberg had made the invention all on his own were Max Knoll and Ernst Ruska. Ruska in particular would stay annoyed about the Rüdenberg patents for his entire life, which is quite understandable even if Rüdenberg's assertions happen to be true. Since Rüdenberg had also been a professor at the Technische Hochschule, Ruska had known him personally, as he told us in 1985:

> *"I was examined as student by Rüdenberg by the end of 1930. I had attended his lectures on "Switch surges and transient effects". That was his speciality. As not many students had attended these lectures, I was the only examinee, and then I was not examined in the Technische Hochschule, but at his home, here nearby, in the Douglasstrasse."* [13]

This street was in the Berlin area of Grünewald, not far from the Max-Eyth-Strasse in Berlin-Dahlem, where Ruska lived when we interviewed him. The exam in 1930 was only a few months before Ruska constructed his first prototype of the electron microscope. Once more this history shows that the very beginnings of electron microscopy are set in a very small world.

The patents only started to become an issue, however, in 1960, after the *Encyclopaedia of Chemistry* [sic; in fact this was the *Encyclopedia of Microscopy*; PWH] had received a letter from Rüdenberg in which he informed the editor that he was the sole inventor of the electron microscope. Upon learning this, the editor asked Knoll what he thought of it. Triggered by the request, Knoll wrote on 17 October 1960 a letter to Rüdenberg's former assistant Steenbeck, who happened to be a friend of his (Mulvey, 1973), to hear what he had to say about it. The reason for Knoll to do this was the fact he and Ruska had shown their new device and the results obtained to many visiting scientists in the period between Ruska's first two stage images of 7 April 1931 and Knoll's Cranz Colloquium of 4 June. Steenbeck had been one of those visitors, and according to Mulvey he had even been given a preview of the colloquium itself then. These visits of scientists have also been reported by Martin Freundlich, one of the fellow doctoral students in Knoll's group. In his reply of 8 November 1960, Steenbeck confirms his visit to the High Tension Lab and he gives Knoll a quite detailed description of what he had seen, after which he writes:

> *"I reported to Rüdenberg very soon after my visit what I had seen in your laboratory and my above-mentioned impression of the implications of your work for cathode ray oscilloscopes, and I certainly talked about the convincing verification of the*

[13] John van Gorkom, Bart de Haas, *Interview with Ernst Ruska in Berlin on 25 June, 27 June and 3 July 1985*, track 4B at 21 minutes 55 seconds.

> *Busch lens formula. Rüdenberg,* not *hampered by the absence of experimental* experience *and hearing of this relation, will have made the following conjecture for himself: if magnetic coils and electrostatic fields work as lenses, as has been shown, then the known kinds of optical instruments can presumably be imitated with electron rays. He thus applied for a patent for this idea, without me knowing about it. If he had spoken to me about it, I would have found the suggestion nonsensical at that time anyway.*
> *Rüdenberg's application is certainly a consequence of my visit to you and was certainly stimulated by my report on what I had seen. The conclusion that as a result of your experiments an electron microscope was possible or at least conceivable, was certainly drawn independently by Rüdenberg himself, thanks to his well-known habit of generalizing. It therefore seems conceivable to me that Herr Rüdenberg really regards himself as the legitimate inventor of the electron microscope, even though he did not make* any experiments or serious computations *for the realization of such an instrument."* [14]

This can be interpreted as a rather damaging testimony, and so to some extent it is surprising that this letter by Steenbeck has not been criticised at all by the authors who paid closer attention to it in later years (Ruska, 1984, 1986; Hawkes, 1985; Freundlich, 1994; Lin, 1995). It seems reasonable, though, to take into consideration that Steenbeck wrote his testimony nearly 30 years after he made his visit to Ruska and Knoll. In his answer, Steenbeck—who spent more than ten years in the Soviet Union after World War II—remarks that he had to go by his memory and that a lot had happened since. One may wonder how accurate his memory can reasonably be. Perhaps more relevant, however, is the fact that Steenbeck lived and worked in Jena, when he wrote his letter, which lay in former communist East Germany in 1960, while Knoll was living and working in Munich, in capitalist West Germany. This was in the midst of the Cold War, General Eisenhower was president of the USA, Khrushchev was ruling the USSR, it was just a year before the Berlin Wall was erected and two years before the Cuban missile crisis nearly triggered a third world war. There was no free traffic of anyone or anything whatsoever between the communist East and the capitalist West and this is what makes this correspondence between Knoll and Steenbeck slightly suspicious. In 1960, Rüdenberg was already for many years a US citizen and by then acknowledged in the USA as

[14] Max Steenbeck to Max Knoll, letter of 8 November 1960, in: Ernst Ruska, The Emergence of the Electron Microscope (Connection between realization and first patent application, Documents of an invention), *Journal of Ultrastructure and Molecular Structure Research* 95 (1986) 3–28, p. 22. Emphases have been used to mark my changes to the original translation. [Also published in Hawkes, 1985.]

the inventor of the electron microscope.[15] Is it inconceivable that in 1960 East-German authorities will have actively encouraged Steenbeck to make damaging statements on an American inventor? Of course, it will be very difficult to prove or disprove this, but maybe worthwhile for someone to examine more closely.

3.4 A Political Dimension

As far as actual electron optical experiments at Siemens are concerned, it is not certain that absolutely nothing was done. In the 1945 draft of his memoir, Rüdenberg wrote:

> *"I asked two of my assistants, Dr. G. Berthold and Dr. M. Steenbeck, to set up in their laboratory an experiment in principle in order to prove the correctness of my deductions, particularly with respect to the imaging effect of an electrostatic field acting like a lens on electron rays. Although they personally did not believe in all of my deductions at that time, they started to build a model. However, some time later they reported that they did not succeed since their device broke down before any test of the principle was made and that hence they had given up the matter in their S.S.W. laboratory. No fresh start was attempted there. In the meantime, one of the Siemens & Halske laboratories had built a small complete electron microscope according to the lines of my patent application" (Rüdenberg, 2010).*

SSW stands for Siemens-Schuckert Werke, Siemens & Halske was a sister firm of SSW. The Siemens-Schuckert Werke were specialised in high-power engineering, while Siemens & Halske covered the field of low-power engineering.

The claim that a microscope had been built at Siemens & Halske has actually been confirmed by a witness at the 1947 law suit in Boston, Massachusetts. In 1931 there had been a young electrical engineer in the Siemens patent department, whose name was Kurt Abraham, and who had fled to the US in 1937, where he changed his name to Curt Avery. As Avery had been immediately involved in drawing up the patent applications in Berlin, he had been invited to testify in Rüdenberg's American law suit. He told the court:

> *"Well, Doctor Lübcke had a laboratory within the research laboratories, engaging in the development of a gas-filled electronic tube for generating high-frequency, intended for medical purposes, and he had at his disposal several glass blowers of exceptional skill, because his particular line required glass-blowing to a very large extent. And when he received a copy [of the application], well, actually I think, I don't know when he received the copy—when I came to him about two or three months*

[15] See for example Rüdenberg's obituary in *Physics Today* 15 (April 1962) 106.

later I saw in his laboratory an electron microscope. It consisted predominantly of glass, with two electro-magnetic coils or lenses, and lots of conduits leading into and out of the structure, and connected with a pump that was running continuously during the operation; and he explained to me that he was trying to find out whether the concepts disclosed in these applications would work" (Rudenberg & Rudenberg, 2010).

This is the same Ernst Lübcke who has been named in this section as one of the other inventors mentioned on electron microscope patents. The moment that Avery visited his lab must have been somewhere in August 1931.

From Avery's testimony it can be learned that Lübcke was forbidden by the Siemens & Halske Board to publish any of his results. Also Rüdenberg asserts that his superiors had forbidden him to publish:

"I had prepared an early description in a scientific journal of the new principles incorporated in the electron microscope on the basis of my disclosures. However, the executives of the Siemens firms objected to such a publication. Only a short note without any details was conceded in 1932. One of their reasons apparently was that they did not want to be connected with a mere fantasy. This was also indicated by the refusal of any further development of the invention in their appertaining departments. Under the reign of the Nazi philosophy at that time I could only yield to their decisions" (Rüdenberg, 2010).

The "short note without any details" is the one in *Die Naturwissenschaften* that has been discussed at the beginning of this section. The Nazi regime is a recurrent theme in Rüdenberg's presentation of the history of the electron microscope, as mentioned before. Already in his very first public claim on the inventorship in his letter to the *Journal of Applied Physics* in 1943 he wrote:

"The development of the electron microscope was continued by other investigators within and without the corporation mentioned above and its affiliated firms, and has lead to great success. Your readers probably know that in Naziland no credit is given to inventors who emigrate."

Taking the historical facts into consideration, it does not seem justified to attribute the unwillingness of Siemens to develop electron microscopes to Nazi politics. Already during the law suit of 1947 it became clear that the Siemens board had gone through considerable efforts to get Rüdenberg and his family safely out of the country in 1936. This has been extensively acknowledged by the Rudenberg family as well in their 2010 comment. Also the time frame is incorrect. The political situation started to become really dangerous for Jewish people from the moment that Hitler came to power. This happened on 24 March 1933 when the so-called Enabling Act passed all legislative powers from the *Reichstag* (the German parliament) to Hitler's government, which had just been formed. After this Act came into force, Hitler could draw up any other act that he wished, effectively meaning the beginning of his dictatorship. The Rüdenberg patents were filed in 1931, so this is two years before Hitler took over.

One of Hitler's new acts was the "Law for the Restoration of the Professional Civil Service", passed on 7 April 1933. As a consequence of this law, anyone who was Jewish for fifty percent or more was not allowed to be employed by public bodies. In this respect, it is certainly illuminating that the Jewish Gustav Hertz, a nephew of Heinrich Hertz, was invited to come to work for Siemens after the Technische Hochschule had been forced to send him away because of this new law. Together with James Franck, Hertz had received the 1925 Nobel Prize in Physics "for their discovery of the laws governing the impact of an electron upon an atom." As late as 1 July 1935, when Nazi politics had become a lot grimmer, Hertz was hired by Siemens[16] where he got his own "Research Lab 2". Altogether, it is clear that the Siemens Board should not be accused of being active supporters of Nazi practices.

A far more likely explanation for the unwillingness of Siemens to develop electron microscopes is suggested by Falk Müller. As the Rudenberg family also has noted in their comment, Müller remarks that there existed an informal agreement between Siemens and AEG according to which AEG was granted the development of electron optics (Müller, 2009). In other words, there seems to have been a cartel. A confirmation of this observation can be found in an article from 1945 by Miletus L. Flaningam in *The American Journal of Economics and Sociology*, in which a civil suit is discussed that had been instituted in January 1945 by the US government against the General Electric Company (GE) and the International General Electric Company (IGE). It explains that:

> *"The Department of Justice charges these companies with alleged conspiracies in violation of the Sherman Act and the Wilson Tariff Act by reason of certain contracts which provide the exchange of patent rights, technical knowledge and skills, manufacturing experience, and for territorial division of the world for purposes of eliminating trade competition" (Flaningam, 1945).*

AEG is closely affiliated with these firms, as has already been mentioned in Section 2. This whole issue was not limited to GE, however. Flaningam

[16] Letter of 25 July 1985 from the Siemens personnel department to Ernst Ruska. Ruska's enquiry after Hertz had been triggered by our discussions with him during the 1985 interview.

pays close attention to the links between AEG and Siemens & Halske, stating for example:

> *"In January, 1930, Siemens–Halske was authorized to issue Rm 30 million in participating gold debenture bonds (...) of which ten million were acquired by GE in February, 1930, at 230 per cent, which represented an investment of twenty-three million dollars.*
> *Although IGE was allowed no voting power in Siemens–Halske, as a result of the above and other financial transaction, the former did share substantially in the profits of Siemens–Halske."*

Flaningam also argues that international agreements were negotiated between Siemens, GE and AEG in order to divide the world market amongst them and eliminate competition. If all this is true, it appears quite likely that Siemens did indeed not want to move into the field of electron optics, since it had been given to AEG, at least in 1931.

As far as Rüdenberg is concerned, it should be concluded that he had not performed any electron optical experiments before he applied for the electron microscope patents, and afterwards neither he nor Lübcke got another opportunity to follow up further developments. This state of affairs was very directly expressed by Martin Freundlich, who contributed in 1963 a paper to *Science*, called "Origin of the Electron Microscope". There he wrote bluntly:

> *"Actually, Rüdenberg, though the first to apply for patent rights, did not contribute directly or indirectly to the early development of the electron microscope."*

This was just a little more than a year after Rüdenberg had died on 25 December 1961, and it resulted in an angry letter to him from Gunther Rudenberg. Freundlich answered that letter politely, but did not feel restrained from adding:

> *"I have no quarrel with the habit of patenting ideas as quickly as possible. That was done often by big companies in Germany. But the scientific value of such a paper does not compare with the work of Knoll and Ruska who proceeded in painstaking experiments before they published their results. Your father has not contributed anything to their development, which led them step by step to an electron microscope."*[17]

From the same letter it can also be learned that his paper in *Science* had been triggered by an obituary in *Physics Today* which called Rüdenberg

[17] H. Gunther Rudenberg, letter to Martin M. Freundlich, 8 May 1963, copy from Ernst Ruska's personal archive; Martin M. Freundlich, letter to H. Gunther Rudenberg, 23 May 1963, copy from Ernst Ruska's personal archive.

"inventor of the electron microscope" without any reserves.[18] It illustrates that it was Freundlich's primary intention to set the priority record right.

At this stage it is necessary to add one detail, which is rather significant, since it completely neutralises the quite poignant political dimension of the priority issue. Martin Freundlich happened to be a Jewish refugee as well. He had been forced to flee Germany in 1934 and, just like Rüdenberg, he had chosen to start a new life as a US citizen. Freundlich, however, never attached political connotations to the priority issue, and as a matter of fact he stayed in touch with Ernst Ruska for more than fifty years—at least until 1985.[19]

3.5 Prior Art

From Freundlich's answer to Gunther Rudenberg it also becomes clear that he was already in possession of a copy of the letter in which Max Steenbeck confirms the suspicion that Rüdenberg had acted on prior knowledge. In his 1963 paper, however, Freundlich did not make any reference to this testimony. He is the first though to draw attention to the fact that Rüdenberg had even attended Knoll's Cranz Colloquium on 4 June 1931, but had not taken part in the discussion. This little suggestive detail has been repeated many times ever since.

Freundlich's commitment to the whole affair is made even more explicit by a second article on the history of the electron microscope, which he wrote more than thirty years later. Again he pays much attention to Rüdenberg's claim and concludes:

> *"It is clear that Rüdenberg applied for his patent only after Steenbeck's reports about the experiments he had witnessed at the High Tension Laboratory" (Freundlich, 1994).*

He draws this conclusion from the mere observation that Steenbeck had been amongst the many visiting scientists who had seen Ruska's first prototype before any patent had been filed. This time, Freundlich does mention the Steenbeck letter, but does not use it as evidence. In a way, Freundlich's conclusion summarises the essence of the whole Rüdenberg story: Rüdenberg was not actively developing any kind of electron optical device, but he

[18] Editorial, Obituaries, *Physics Today* 15 (April 1962) 106.

[19] In September 1985 I received a letter from Martin Freundlich in which he forwarded a letter that Ernst Ruska had intended to send to me together with a copy of the lecture by his father, mentioned in Section 2. Ruska had sent the article to Freundlich as well, while the letter to me had accidentally ended up in Freundlich's envelope.

was brilliant enough to realise the potential of such a device, once he had become aware a functioning one had been constructed. [A paragraph has been omitted here as not appropriate for this discussion; PWH.]

4. LÉO SZILÁRD

Already twice in this part there has been a reference to Dennis Gabor's representation of history according to which the world of physics was thinking of electron microscopes and telescopes in as early as 1927, immediately after Hans Busch had shown that you can have electron lenses. Gabor made his remarks in 1942 and that was not the last time that he liked to stress how obvious electron optics had actually been. In 1953 he presented the same view in a somewhat different way, telling his audience at a conference in Bournemouth:

> *"Let us look back at the year 1928. The publications of Hans Busch, which have become the foundation of geometric electron optics, had appeared just a year before, and wave mechanics had become part of the general education of physicists. How then could it happen that anyone who was able to add up two and two, would not have thought of an electron microscope, and how is it possible that he would not have been thrilled by its brilliant prospects? These are difficult questions. Instead of trying to answer them, I will reproduce a conversation, which took place in a Berlin pub at the beginning of 1928 between my friend L. Szilárd and me, and which I remember, I believe, nearly word by word. Szilárd: 'You know, Busch has shown that you can have electron lenses. You have the skills, why don't you make a microscope with electrons? You would be able to go down to the de Broglie wavelength!' (...) (Gabor, 1957).*

And twenty years later, in 1974, at the Eighth International Congress on Electron Microscopy in Canberra, Australia, Gabor expressed his disbelief about the fact that Hans Busch had not suggested any kind of electron optical application in his famous paper of 1927, after which he continued to say:

> *"There were people with more imagination than Hans Busch, in particular my friend Léo Szilárd, probably one of the brightest inventive minds that ever existed. It was from him that I first heard the words 'electron microscope.' I think I remember a conversation which I had with him in 1928, in the Café Wien, in Berlin, almost verbatim. Szilárd: 'Busch has shown that one can make electron lenses, de Broglie has shown that they have sub-Ångström wavelengths. Why don't you make an electron microscope, one could see atoms with it!' (...)" (Gabor, 1974).*

As we see, the words attributed to Szilárd changed considerably over the years, so at least one version is not as verbatim as Gabor believed himself.

It should be especially noted that the words "electron microscope" were not used in the first quote, where it is called a "microscope with electrons." Nevertheless, it is clear that Szilárd might very well have been the first to utter the idea of an electron microscope.[20]

We may wonder, however, how much value should be ascribed to such a claim. After all, we cannot know whether there has been a similar remark by any other bright mind at the time. Would for instance de Broglie not have had a fleeting thought about possible applications,[21] followed by an accidental remark, when he learned of Busch's electron lens? Gabor believed that Hans Busch himself did not think of applications, but it is not unlikely that Busch was simply unwilling to make any public claims as long as his theory had not been properly verified yet. Is it certain that he never made an off-the-record remark about theoretical applications of geometrical optics with electrons? For this reason, there would have been little justification to dedicate a whole section to Szilárd, if there had not existed additional evidence for the special interest that he took in the electron microscope.

Although Szilárd gained fame as a nuclear physicist in later years, in the early 1920s he had also studied for some time at Berlin's Technische Hochschule (Feld, 2008; Frank, 2008). Maybe this engineering background explains that he decided to apply for a patent on an electron microscope in 1931. The German patent 965 522 that he obtained is simply called *Mikroskop*. Unfortunately for him, the priority date is 4 July 1931, so exactly five weeks after the first Siemens–Rüdenberg patents. As a consequence, the published patent contains a footnote that there exist older rights, referring to the German Siemens patents 889 660 and 895 635. In comparison to those patents, the Szilárd patent contains two remarkable novelties, however. In the first place, it also covers the use of charged particles other than electrons. As the very short patent contains no theoretical details whatsoever, there is no reference to de Broglie's equation to underpin this claim. Siemens was going to patent the idea of using other charged particles nearly a year later, as has already been pointed out in the previous section. Szilárd was also the first to come up with the idea of further enlarging the electron microscopical image by means of a light microscope. In this case too, Siemens included the idea in a later patent.

[20] A third place where Gabor claims that already in 1928 Szilárd had coined the words electron microscope is in his preface to Ladislaus L. Marton, *Early History of Electron Microscopy*. San Francisco: San Francisco Press, 1968, p. VI.

[21] Louis de Broglie did indeed suggest electron optics as a subject for a PhD to a student in 1927 or 1928 but the latter preferred a different subject (de Broglie, 1946); PWH.

Szilárd did not use the word electron microscope, nor does he claim to have invented a microscope that would have a better resolution than a light microscope, although he does point at the principal limitations of light microscopy. In a way it is fortunate that he did not claim a superior resolution, since the accompanying drawing presents a device that would certainly have failed to do so. It contains three magnetic coils, positioned outside the cathode ray tube's wall, outside the vacuum. In other words, the inner diameter of the coils is so wide that the complete cathode ray tube itself can pass through it. Furthermore, the coils do not appear to be encased in any sort of way, nor is there any reference to such an encasement in the text. This means that the lenses have very large focal distances, and the resulting magnetic fields are substantially weakened by the tube's wall, which has to be penetrated before the fields reach the rays that have to be focussed. Altogether, the performance of this microscope would be very poor—as a matter of fact, just as poor as Ruska's first prototype which also applied external coils without caging.

Maybe the most intriguing aspect of this Szilárd patent is its date. If he had already thought of an electron microscope in 1928, why did he wait for three years until after Knoll had made Ruska's results public at the Cranz colloquium? What bothers one most here is probably the fact that neither the Knoll group, nor the electron physicists at the AEG institute nor a man like Rüdenberg had managed to grasp the significance of geometric electron optics before 1931. In that light you may wonder whether it is really a coincidence that Szilárd too started to act in 1931, at the same time as his competitors. Maybe the date of 1928, as Gabor remembered it, is simply incorrect.

5. ELECTRON OPTICS IN THE USA AND GREAT BRITAIN

So far, the invention of the electron microscope has proven to be solely a Berlin affair. All four competing parties were based in Berlin—the major three even within walking distance from each other. These major three—Technische Hochschule, AEG and Siemens—all applied a technical approach in which the cathode ray tube was the point of departure and the implementation of the Busch lens theory the challenge. There is hardly any mention of the fresh fruits of new theoretical physics, let alone that such modern insights would have been a source of inspiration to them.

In contrast to this, it seems that English and American physicists travelled the road to electron optics in exactly the opposite direction. After de Broglie had formulated the particle–wave nature of electrons, in 1927 American and British physicists would demonstrate that he was right. In America this was done by Clinton J. Davisson and Lester H. Germer of Bell Telephone Laboratories in New York, and in England by George Paget Thomson of Aberdeen University—the son of J.J. Thomson, the man who discovered the electron. Both Davisson and Thomson received a Nobel Prize in Physics for it in 1937. After having established that rays of electrons behave like waves, no doubt it was a logical step to start thinking about constructing some kind of lens. On 1 August 1931, the *Physical Review* of the American Physical Society published an abstract titled "Electron lenses", written by Davisson and Calbick (1931, 1932). Their mathematical investigation lead them to the conclusion that: "The field about a circular hole in a plate is the analogue of a spherical lens." Approaching the matter from a completely different angle, they had arrived at the same point as Hans Busch had done four years before. This 'North-Atlantic route' in history might explain, by the way, why it is a widespread belief that the conception of the electron microscope started with de Broglie.

In 1931 also, George Paget Thomson's main interest was a different one. On 4 December that year, he delivered a Friday evening discourse at the Royal Institution in London, titled "Electron Optics", a topic which was just as new in London then, as it was in New York or Berlin. The emphasis of his talk lay, however, on de Broglie's recent wave–particle equation and the new field of electron diffraction—another field of optics, which he himself, as well as Davisson and Calbick, had applied four years earlier to show that electrons do indeed behave like waves. Diffraction is the result of interference, which is the phenomenon that waves can merge. If for example, the crests of one wave coincide with the crests of another one, the crests will fuse—two smaller waves add up to one bigger wave. However, if the crests of one wave happen to coincide with the troughs of another wave, the crests and troughs will extinguish each other—the two waves will annihilate each other. This was shown to be possible with rays of electrons too, and the extinction is especially strange as soon as you imagine electrons to be particles. As Thomson said in his lecture about the study of electron diffraction patterns:

> *"While these experiments show the close resemblance between the behaviour of the electrons and that of light in those experiments which were the original proof of its wave character, yet, all the same, one must not entirely abandon the particle idea. When the electron reaches the photographic plate, it seems to forget that it is a wave and becomes a particle. If the plate were examined under a high magnification, the pattern would show up as a number of black specks, each corresponding*

> *to one grain of silver bromide made developable. Each electron affects one grain only (. . .).*
> *On the wave side of its nature the electron is a widely extended entity which in a sense occupies the whole region in which it might be found. An electron is like an able guerrilla leader who occupies a wide area with rumours of his presence, but when he strikes, he strikes with his whole force" (Thomson, 1931).*

Thomson saw a practical use of electron diffraction for the study of the arrangement of atoms in crystals. He had for example obtained a diffraction pattern formed by electrons that were reflected from a cube face of a silver crystal. Such a pattern reveals information about the arrangement of the atoms at the surface. Thomson, however, did not say anything about geometric electron optics on 4 December 1931, and made no reference whatsoever to possible electron optical instruments.

An exception to the North Atlantic route in history might be found in the person of Paul A. Anderson, a US physicist who had learned about Knoll and Ruska's work during a stay at the *Physikalisch-Technische Reichsanstalt* in Berlin. In 1987, he wrote in a letter to electron microscopy pioneer John Reisner:

> *"In 1930, after spending the first year of a National Research Council Fellowship at Harvard, I went to the Low Temperature Laboratory of the P.T. Reichsanstalt in Berlin to carry out some work-function measurements on tungsten at liquid helium temperatures. The tube design involved the electrostatic focusing of an electron beam on a distant metal target, and so a self-taught course in electron optics. In going through the literature, I ran across the Ruska–Knoll papers and was impressed... I did not take time to visit their laboratory, an omission I have ever regretted" (Reisner, 1989).*

He subsequently tells that he returned to the US in August 1931. This means that he could have seen Ruska's very first article with Knoll, which had just been published in the August edition of the *Zeitschrift für technische Physik*. However, this first article was not about a microscope yet, so we have to assume that Anderson kept himself informed about later developments after his return to Washington State University, where he was to be appointed as professor. Otherwise he must have been mistaken about the exact dates. Eventually, the work by Knoll and Ruska inspired the construction of a magnetic transmission electron microscope, as Anderson described in the same letter:

> *"(. . .) Kenneth [Fitzsimmons] asked me to suggest a research program to work on. I suggested several projects suited to our sources, but he had heard me talk about Ruska's work, and it was evident that this heart was set on building a microscope."*

It would take until December 1937 before Anderson and Fitzsimmons managed to make the first American electron microscopical images with their instrument.

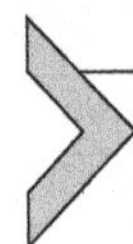

6. CONCEIVING THE IDEA: SUMMARY AND CONCLUSION

In order to obtain a better overview of the developments that took place between 1928 and 1932, I will start this concluding section with a chronological summary of the major events and achievements between 1928 and 1932.

Year 1928 Léo Szilárd is believed to have made a casual reference to the concept of an electron microscope (Gabor, 1957, 1974).

10 May 1929 Ernst Ruska completes his student research project (*Studienarbeit*) in which he confirmed Busch's lens theory. He also paid attention for the first time to the concept of an iron casing as a means to concentrate the magnetic field. The advantage would be a reduction of the current needed in the coil, he argued at the time (Ruska, 1929).

Year 1930 Ernst Brüche writes in an AEG book a chapter about alternative applications of the cathode ray tube and is the first to use the terms "geometric electron optics" and "electron lens" in a publication (Brüche, 1930).

23 December 1930 Ernst Ruska completes his Master thesis (*Diplomaufgabe*) in which he made remarks about the observation of spontaneous magnifications and of scattering (Ruska, 1930).

7 April 1931 Ernst Ruska makes his first two-stage images with a device that can be described as the first prototype of the electron microscope (Ruska, 1979).

28 April 1931 Submission date of the first joint article by Ruska and Knoll, in which they presented Ruska's confirmation of the lens theory. They also paid renewed attention to the iron casing as a means to concentrate the magnetic field. This time the advantage of the casing would be a reduction of the number of electrical windings in the coil, resulting in a reduction of weight (Ruska & Knoll, 1931).

30 May 1931 Priority date of the Siemens–Rüdenberg patents that were applied for in Austria, France, Great Britain, Netherlands, Switzerland

and USA. These eight patents cover the same claims as the 14 German ones that are mentioned below (see Appendix A).

31 May 1931 Priority date of the German Siemens–Rüdenberg patents 889 660, 895 635 and 906 737, and the additional German Siemens patents without inventor 754 259 and 758 391. The patents cover the design of an electrostatic lens and the concept of using electron optics to concentrate the electron beam, to image objects and to magnify objects. As such, three of them describe an electron microscope. Reference was made to chromatic aberration and to the imaging of objects by means of an electron beam that passes through them. The term "objective lens" was also mentioned. The terms "electron microscope" and "chromatic aberration" were not used as such yet (see Appendix A).

4 June 1931 Knoll delivers the first public presentation of Ruska's results in which he likened the new device to a telescope and not to a microscope (Knoll, 1931).

27 June 1931 Priority date of German Siemens–Rüdenberg patent 916 838, a supplemental patent to cover the idea of making magnifications by means of electron rays (see Appendix A).

28 June 1931 Priority date of German Siemens–Rüdenberg patent 911 996, a supplemental patent to cover the idea of making magnifications by means of electron rays with different speeds (see Appendix A).

4 July 1931 Priority date of Szilárd's German patent 965 522. He mentioned the use of small particles other than electrons, and the idea of adding an optional light optical stage to the electron optical magnification stages (Szilárd, 1931).

1 August 1931 Davisson and Calbick publish a short abstract titled "Electron lenses" in the *Physical Review* of the American Physical Society (Davisson & Calbick, 1931).

10 September 1931 Submission date of the second joint paper by Knoll and Ruska. It is the first time that the terms "electron microscope" and "chromatic aberration" were submitted to print, and also the first time that the term "objective lens" actually appeared in print. Once more they paid attention to the iron casing as a means to concentrate the magnetic field, and this time they noticed that it would shorten the focal distance. Subsequently they made a calculation to demonstrate that a magnification of 3375 times would be feasible. They mention the possibility of imaging very small objects and explain that intense electron beams are needed for high magnifications (Knoll & Ruska, 1932b).

Month of November 1931 Brüche submits his short notice called *Elektronenmikroskop* to *Die Naturwissenschaften*, in which he claims that AEG is actively involved in the construction of electron microscopes as well. It is the first time that the term "electron microscope" appears in print (Brüche, 1932a).

31 March 1932 Priority date of the German Siemens–Rüdenberg patents 916 839 and 916 841, both labelled as a supplement to 916 838 and both dealing with the magnified imaging of objects by means of electron rays (see Appendix A).

15 May 1932 Priority date of German Siemens patent 909 156, which mentions no inventor, but also describes a device to magnify images by means of electron rays (see Appendix A).

7 June 1932 Rüdenberg submits his short notice called *Elektronenmikroskop* to *Die Naturwissenschaften* in which he claims that Siemens is actively involved in the construction of electron microscopes as well (Rüdenberg, 1932).

18 June 1932 Priority date of the German Siemens–Swinne patents 915 843 and 916 840 of which the latter is labelled as supplement to the Siemens–Rüdenberg patent 916 838. Both patents cover the magnified imaging of objects (see Appendix A).

13 August 1932 Priority date of German Siemens–Rüdenberg patent 915 253, which deals with the design of an electrostatic lens (see Appendix A).

A closer look at this summary shows first of all that most of the important events took place in 1931. Out of 20 listed dates, 11 lie in 1931, while 4 out of 5 later dates are certainly less significant, since they refer to supplementary Siemens patents. 1931 proves to be the year of the first prototype of the electron microscope, its two-stage images and the first public presentation of these results. 1931 is the year of the decisive Siemens–Rüdenberg patents as well as Szilárd's patent, and the year of the first use of the words "electron microscope", "objective lens" and "chromatic aberration." Finally, 1931 is the year the suggestion was made to substantially magnify small objects with electron rays and to use an iron casing in order to reduce the focal distance of an electromagnetic lens. Therefore, it seems absolutely justified to call 1931 the year that the electron microscope was invented if the reader wishes to pinpoint a date for such an event. At the same time, 1931 is certainly not the year that the significance of de Broglie's equation was realised. Evidently Davisson was familiar with de Broglie, as he had even provided the experimental confirmation of de Broglie's thesis. However, the equation was not mentioned by him and Calbick in their abstract of 1931. Also Szilárd seems to have understood, as could be concluded from his patent, but he did not refer to de Broglie there, and even if he had done so, nobody could have known, since the patent was not published until 1957.

When weighing the contributions of the four rival parties, we immediately see that most emphasis lies on the Technische Hochschule (Knoll and Ruska) on the one hand with six dates, and Siemens (Rüdenberg) at the other hand with nine. The contribution of AEG (Brüche) is limited to coining the terms "electron lens" and "geometric electron optics" in 1930, as well as the name "electron microscope" in November 1931. We may therefore conclude that AEG's contribution was rather a linguistic one at this stage.

Szilárd's credits consist of a patent that was filed five weeks too late and Gabor's assertion that Szilárd had mentioned the idea of an electron microscope already in 1928. Self-evidently, it is very difficult to judge the value of such an assertion made by a single witness, especially since Szilárd only became actively involved—and only in the sense of applying for a patent—*after* the moment at which the other three parties had already entered the field.

All nine dates that refer to Siemens concern in one way or the other patents that were filed, especially by Rüdenberg. Even the Rudenberg family has accepted the conclusion that the ideas laid down in the first patents had not been experimentally verified at all by Reinhold Rüdenberg (Rudenberg & Rudenberg, 2010). Nevertheless, the patents do contain a number of claims and considerations that have relevance to the priority debate. First of all, Rüdenberg was the first to describe the principle of an electron microscope on paper. Maybe we should not give too much weight to the legal notion that he did this in a patent, as the patents expired a long time ago, but that does not take away the historical fact that on 30 May 1931 documents had been drafted which clearly describe an electron microscope—at least, as far as we can judge on the basis of the patents that were finally published. His description was rather detailed insofar that he already coined the term "objective lens" then, and was paying special attention to chromatic aberration, although not yet using that term. On the other hand, where it comes to electrostatic lenses, Rüdenberg also described concepts that were never going to work.

In great contrast to the nine Siemens dates, all six events and achievements that are linked to the Technische Hochschule are dominated by actual experiments which show a steady and continuous progress. This starts in 1929 with the verification of the lens theory, followed by the dismissal of an ill-designed electrostatic lens in 1930, and finally in 1931 the construction of a device that was able to make two step magnifications. Especially fascinating is the steady accumulation of little insights which slowly aggregate into what was to become a single brilliant concept. In 1929, Ruska paid attention to an iron casing and realised that it would concentrate the magnetic field in a smaller space, thereby making it more effective. At that stage, however, the only relevance that he saw was the reduction of necessary current in the coil. In the first paper by Ruska and Knoll, submitted in April 1931, the casing is extended to the inside wall of the coil, and then they mention that this will be helpful to reduce the number of electrical windings in the coil. Four months later, they note in their second joint paper that the advantage of the iron casing will be a reduction of the focal length. So what we see is an apparently deliberate shift of emphasis towards the relevance for microscopy, all based on the same equations. This shift must have been rather conscious, since the equations had not changed in the mean time and therefore Ruska and Knoll knew right from the beginning that focal distance was one of the parameters involved. Their original subject was the oscilloscope, however, and for these instruments a short focal distance is of less importance. Other isolated insights were the observations of spontaneous magnifications and the electron scattering by a very thin foil—both in 1930—and the adjustability of the focal distance, mentioned in April 1931. It all accumulated in the second joint paper where they added the concepts of the objective lens, chromatic aberration, multi-stage enlargements in order to obtain very high magnifications, the idea of imaging very small objects and the need for electron beams with a very high intensity as soon as high magnifications are going to be made. The concept of the electron microscope had clearly been conceived, at least in Ruska's perception. As it is shown in this part, there is little reason to assume that Knoll gave much credit to the idea.

The crucial difference with Rüdenberg is that Knoll and Ruska put experiment in the first place and the description of concepts in the second. Rüdenberg's very first step was to write the concept down, and whether this step was followed by experiments is not even that relevant here. According to Max Knoll, Ernst Ruska, Martin Freundlich and Max Steenbeck, Ruska's prototype of an electron microscope had already been demonstrated to visitors before Rüdenberg wrote his patent applications. Effectively, therefore, the device already existed before Rüdenberg had described it. Nevertheless, Rüdenberg's description on paper is earlier, than the one by Knoll and Ruska. So, going by patents and articles alone, it may look as if Rüdenberg had conceived the electron microscope earlier. An important aspect of this priority matter is also the issue whether Rüdenberg made his own extrapolations after hearing Steenbeck's reports on Ruska's novel contraption, or whether he simply took advantage of the situation. In this respect, at least Steenbeck gives him the benefit of the doubt. Ruska certainly does not.

No doubt, the question who was first to come up with the idea of an electron microscope is by far the biggest priority issue in comparison to two smaller ones. The first of these minor issues is the question who was first to coin the words "electron microscope." If you go by submission date, Knoll and Ruska were first, but if you go by publication date it was Brüche. Another issue could be the question who was first to coin the term "objective lens." It appears for the first time in the Siemens–Rüdenberg patent 895 635 with the priority date of 31 May 1931, but we do not know whether it was also used in the application itself.

To conclude this section, I wish to take a final look at the motives of those involved. Szilárd's motive is probably the most straightforward. He was not actively involved in any related research, so it seems he simply wanted to patent a good idea.

In contrast to Szilárd, Ruska appears to be the classic example of the researcher who gradually becomes aware of the potential of the theories that he is experimentally verifying.

In the case of Knoll one may sincerely wonder whether he ever believed in anything like an electron microscope at this early stage. He likened Ruska's prototype to a telescope, he had not made any efforts to protect their intellectual property, on 13 November 1931 he even informed Brüche about their upcoming paper with an unfortunate priority dispute as a result, and finally he left the Technische Hochschule in 1932 in order to join the nascent television industry.

As far as Rüdenberg is concerned, it was first of all his job as Chief Electrical Engineer of the Siemens-Schuckertwerke to innovate, so that would be the most straightforward motive to be attributed to him. According to himself, it was his son's infantile paralysis that drove him to come up with an electron microscope. As a matter of fact, there is absolutely no reason why the first motive should exclude the second. The credibility of the second motive is poor, however, especially when you take into account the summary of Rüdenberg's remarkable train of thoughts that was presented at the 1947 law suit in Boston.

Probably, Ernst Brüche is the competitor who is most difficult to understand at this stage. It is clear that he did not like the high voltage engineering that is needed for electron microscopes. As will be shown in the following part, his main optical interest was the imaging of cathode surfaces with slow electrons and a single electrostatic lens, so one may even dispute whether Brüche's idea of an electron microscope had anything to do with the instrument that Ruska and Rüdenberg envisioned. For Brüche, the most straightforward motive would be company interests. We already saw that the emerging field of electron optical applications had been granted by Siemens to AEG in 1930. Most likely, Brüche was taking the protection of this new entrepreneurial territory very seriously. This would also explain some of the events yet to come.

PART 2. MATERIALISING THE IDEA

The question of who deserves to be granted the honour of having invented the electron microscope will never be answered to everyone's full satisfaction. It is very clear, however, that the event as such took place in the year 1931, while an appropriate name for the invention quickly followed that same year. In the succeeding two years, the envisioned new instrument would develop from an intriguing concept into tangible apparatuses that were eagerly presented as "electron microscopes" as will be painted in this part. What will interest us most from a historical perspective, is how this label was used in practice and to what extent we are dealing with true forerunners of the instrument that we know today. Closely connected to this matter of definition is the matter of motives again. Whatever the people involved imagined they were constructing, why were they doing it?

In an attempt to answer these two main questions, I will apply a format which is rather similar to the one used in Part 1. In Section 7 we will take a look at the activities of those involved with the Technische Hochschule. In Section 8 the work of Brüche and colleagues at the AEG Research Institute will be discussed. In the third, fourth and fifth sections I will pay attention to the first developments that took place outside Germany during these two years. The characters of Szilárd and Rüdenberg will not reappear in this part since they play no role whatsoever during the period discussed. Nor will I elaborate on the first strictly theoretical studies of electron optics that started to appear in this period (Glaser, 1933a, 1933b, 1933c), with the exception of one or two that are immediately relevant here. In the concluding section (Section 12), answers to the main questions are formulated. These answers will also be linked to the conclusions of Part 1 in order to identify recurrent themes that could be used to draw up the first outlines of a general overview.

Finally, I wish to stress that I will occasionally go to some length to discuss rather specific details, if I consider them to be especially relevant to later priority issues.

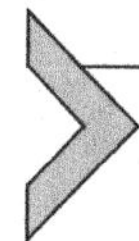

7. TECHNISCHE HOCHSCHULE

7.1 Bodo von Borries

As we saw in Section 1, Ernst Ruska's tutor Max Knoll decided to say farewell to the High Tension Lab and join the development of television at Telefunken's, starting 1 April 1932. His position was taken over by Bodo von Borries, who had been one of the PhD students on his team since 1 April 1929 (Grünewald, 1976; von Borries, 1991; Ruska, 1957, 1970a).[22] On 24 March 1932, a week before he succeeded Knoll, von Borries had completed his thesis about the application of external photography to oscilloscopes. As Ruska explained in two of his historical reflections (Ruska, 1957, 1970a), von Borries had studied the possibility of registering oscillograms on photographic material outside the vacuum, and hence from the outside of the oscilloscope. To do this, he used a so-called Lenard window, which is a metal film that is thin enough to let most of an electron beam pass through, while at the same time it is impermeable to air. The Lenard window replaced the usual fluorescent screen at the end of the cathode-ray tube and the photographic plate was adjusted right behind the Lenard window within reach of the electron beam that penetrated the window.

According to Heinrich Grünewald, von Borries was appointed for four months as private assistant to Professor Adolf Matthias and was urged to hurry to take over Knoll's position. In a letter to von Borries, dated 20 March 1932, Matthias gave as a reason that the High Tension Lab-

[22] I have not been able to find a publication of the PhD thesis by von Borries. In his obituary, published in 1956, the title of the thesis is given as "Außenaufnahme am Kathodenstrahloszillographen", Ernst Ruska, Bodo von Borries (Obituary), *Zeitschrift für wissenschaftliche Mikroskopie* 63 (1957) 129–132. According to Martin Freundlich (1994) it was published in *Forschungsheft #3* of the *Studiengesellschaft [für Höchstspannungsanlagen]*.

oratory was expected to move very soon to a location at the outskirts of Berlin near Neubabelsberg, not far from Potsdam (Grünewald, 1976). Martin Freundlich—one of the other PhD students—has described this same episode in his historical contribution of 1994:

> *"v. Borries became [Knoll's] successor as Professor Matthias' personal assistant. His tasks were to organize the move of the High-Tension Laboratory, except for the one-megavolt transformer, to the Landgut Eule, the former* Kaiser Wilhelm Institut für Sprengstoffforschung, *at Novawes near Neubabelsberg. He was also to supervise the construction of cathode ray oscillographs contracted by Matthias to be built for certain utilities. Thirdly he was to keep an eye on the work of Ruska and Freundlich, the two members of the Knoll Team still working at the T.U. (Freundlich, 1994)."* [23]

Certainly for Ernst Ruska, von Borries' new responsibilities were a pleasant change. Von Borries had joined the Knoll Group exactly three years before, which was just before Ruska completed his first assignment to verify the validity of Busch's lens theory. Later on, after Ruska had come back to the Knoll Group to do his second assignment, they had become good friends. Ruska himself stated in 1970 after receiving the Paul-Ehrlich-and Ludwig-Darmstaedter-Prize:

> *"Our experimental apparatuses were in the same room (...). We often depended on mutual experimental help, and soon a personal friendship developed that went beyond shared scientific interests (...)" (Ruska, 1970a).*

According to von Borries' later wife Hedwig the friendship intensified after a cycling trip to the Baltic Sea, which the two men had made together for a holiday in May 1931. This was just a few weeks before Knoll presented Ruska's first two-stage electron images to the outside world. "From this stage onwards," Mrs von Borries writes (1991), "von Borries and Ruska worked intensively together because of their shared unerring belief in the future of electron microscopy." Evidently it is hard to verify whether this was really their main motive. She published her account in 1991, which was 35 years after the untimely death of her husband in 1956. Hedwig von Borries had been a close witness, however, since she happened to be the youngest sister of Ernst Ruska, living with her brother in the same house at the time. The claim also fits the timeframe given in Section 1 where

[23] "Landgut" means "estate". "Institut für Sprengstoffforschung" means Institute for Research on Explosives. T.U. should be T.H. In 1985, Ruska and his wife had taken Bart de Haas and me on a drive along several places of historical interest. The actual location of the estate was along the Teltow canal, not far from a place called Albrechts-Teerofen. From here, Neubabelsberg itself is several kilometres to the West.

Ruska is quoted as saying that he started to believe in the possibilities of an electron microscope somewhere at the beginning of 1931.

Heinrich Grünewald can also be considered an eye-witness as he had been an assistant to Matthias in 1931. In an article that appeared in 1976, in which he fervently argued that historiography placed too much emphasis on the work of Knoll and Ruska and too little on the roles of von Borries and Matthias, he pinpoints the start of an intense cooperation between Ruska and von Borries at the beginning of 1932. His evidence comes from a note calendar on which von Borries' kept record of the working sessions he and Ruska held in their private time (Grünewald, 1976).[24] It started with three sessions in January and three in February, followed by 11 sessions in March, 11 in April, 6 in May, 9 in June, 19 in August, 20 in September, 21 in October, 7 in November and 12 in December 1932. These meetings were dedicated to activities like "calculating" and "working".[25] The number of entries in 1932 amounts to 122, when going by Grünewald. Hedwig von Borries touches on the same subject and speaks of a total of 97 "evenings and weekends", which is still approximately twice a week the whole year through (von Borries, 1991).

The earliest fruits of their zeal are easy to find: on 17 March 1932 two patent applications were filed in the names of von Borries and Ruska together. The first of the two was titled *Magnetische Sammelspule kurzer Feldlänge* (Magnetic concentrating coil with short field length).[26] It described the encased electron lens which would later become known as polepiece lens—the most elementary component of all future magnetic electron microscopes.

The idea for this lens, which is shown in Fig. 4, was an elaboration on the original iron-cased coil with the concentric circular gap at the inside of the coil. At the place of the gap they had projected a narrowing of the passage through which the electron beam passes. As a result of this the magnetic field will be further intensified, creating a stronger lens with a shorter focal distance. The elements that concentrate the field inside the passage are called polepieces, hence the name. Apart from its historical

[24] It is unclear why Grünewald mentions at the end of his paper (halfway down page 222) a professional cooperation between Ruska and von Borries in 1931. This year is not supported by his own arguments.

[25] In German it says "Arbeiten," which could also mean "articles" here. However, in connection to the use of the entry "calculating" the meaning "working" is more likely.

[26] Bodo von Borries and Ernst Ruska, Magnetische Sammellinse kurzer Feldlänge, *German Patent 680284*, priority date of 17 March 1932, published on 3 August 1939.

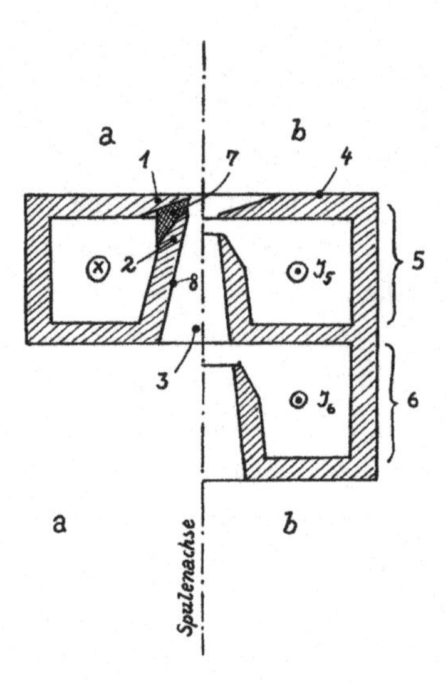

Figure 4 Three different ways to dimension polepieces. The path of the electron beam is vertical, coming from the top. From von Borries and Ruska (1932b).

significance, this patent is a loud and clear expression of the desire to build an instrument that will produce large magnifications, as the two inventors explained themselves right at the start of the patent's introduction:

> *"In the case of electron optical imaging, for example with an electron microscope, large magnifications will in general require the construction of an apparatus of considerable, mostly undesirable length. To avoid this handicap, it is necessary to have small object distances and correspondingly small focal distances of the electron lenses, particularly the objective lens."*[27]

Their second patent was of a less fundamental nature, but it is just as perfect an expression of their wish to produce large magnifications. It is titled *Anordnung zur Beobachtung und Kontrolle der im Strahlengang eines Elektronenmikroskops mit zwei oder mehr elektronenoptischen Vergrösserungsstufen auftretenden elektronenoptischen Bilder* (Device to view and control electron optical images that appear in the trajectory of an electron microscope with two or more electron optical magnification stages) and describes what would later become known as the intermediate screen.[28] The goal of such a screen was

[27] Bodo von Borries and Ernst Ruska, Magnetische Sammellinse kurzer Feldlänge, *German Patent 680284*, priority date of 17 March 1932, published on 3 August 1939, lines 1 to 9.

[28] Bodo von Borries and Ernst Ruska, Anordnung zur Beobachtung und Kontrolle der im Strahlengang eines Elektronenmikroskops mit zwei oder mehr elektronenoptischen

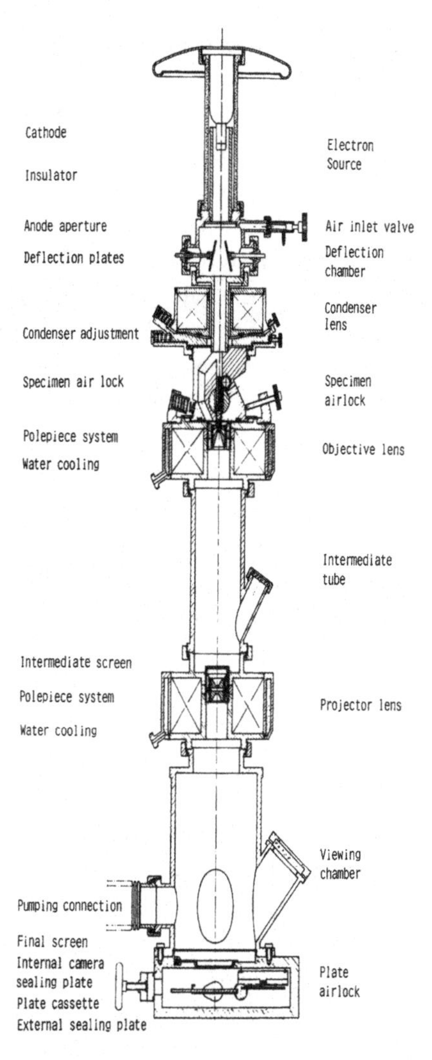

Figure 5 Schematic diagram of the first Siemens commercial electron microscope showing the intermediate screen. After Ruska (1980). Courtesy Hirzel Verlag.

to determine the magnification and also to have a better overview of what you are looking at. The basic idea was very simple and yet very effective. In a microscope, the first lens of the microscope projects its image on top of the second lens, as can be seen in Fig. 5, which shows the first commercial

Vergrösserungsstufen auftretenden elektronenoptischen Bilder, *German Patent 679857*, priority date of 17 March 1932, published on 15 August 1939.

microscope design in which the screen was employed. Only a small part of this first image will fall through the narrow centre of the second lens and be magnified in the second stage. The largest part of the first image is projected on top of the casing of the second lens, so if you put a viewing screen there, you can actually see the first stage image.

The intermediate screen was supposed to have big advantages. As the first stage may give a magnification of a 100 times, one will get a good overall image of the object. The intermediate screen also enables one to make a proper estimation of the size of the total two-stage magnification. It was simply a matter of carefully measuring the diameter of the circular opening in the object holder and the diameter of its projection on the first screen. This would give the size of the first magnification. Subsequently one had to know the diameter of the hole in the second lens and the diameter of its image on the second screen. This would give the second magnification. The multiplication of these two values results in the final magnification. In later years, the intermediate screen has been present in numerous instruments. It should be recalled that at the time of filing this patent not a single electron image of significant magnification had ever been made yet.

The desire to apply for these two patents stemmed at least in part from concerns about rumours that there were competitors with the same intentions. At that moment—in March 1932—they did not know yet about the patent that Reinhold Rüdenberg had filed on behalf of Siemens; it was only three and half months later that Rüdenberg's short notice appeared in *Die Naturwissenschaften.* Ruska paid attention to this subject in his historical contributions from 1984 and 1986, but all that he said is:

> *"Even before Rüdenberg's patent applications were known, the author [i.e. Ruska] had understood that Siemens had applied for patents for the electron microscope" (Ruska, 1984, 1986).*

Whatever it was, the threat that a competitor might run away with their ideas, was quite upsetting, as can be learned from a letter that Ruska wrote about the patent applications to Professor Matthias on 16 March 1932:

> *"Because of your journey, we did not manage to get your consent for the application, despite several attempts. As even a postponement till tomorrow might cause the loss of the priority, we were facing the necessity, either to completely give up the whole enterprise with great probability, or to file the patent at our own cost and risk without your approval" (Ruska, 1979, 1986).*

Five weeks later, on 22 April 1932, Ruska and von Borries submitted their first article together to the *Zeitschrift für Physik* (von Borries & Ruska,

1932a). It was called *Das kurze Raumladungsfeld einer Hilfsentladung als Sammellinse für Kathodenstrahlen* (The short space charge field of an auxiliary discharge as concentrating lens for cathode rays). According to a footnote, the paper was based on a talk at the Technische Hochschule that had been given by Ruska on 27 January 1932. It was presented as a preliminary report and it discussed the idea of using ionized gas as a lens as the title already tells us. This idea had been inspired by the work of Brüche and others on thread beams (*Fadenstrahlen*). As explained in Section 2, a thread beam is held together by traces of ionised gas along its complete length, and so the action of the ionised gas can be likened to a long electromagnetic coil in the way Wiechert applied one in 1899 for the first time (Wiechert, 1899). Von Borries and Ruska describe a method to apply the space charge only locally, in that way creating a concentrating effect that could be likened to a short coil, i.e. an electron lens, which they coined "space charge lens." In contrast to the two patents, it does not seem really justified to label the article as a contribution to the development of electron microscopes. Although it has a type of electron lens as main subject, it was published as a paper from the High Tension Lab, the authors thank professor Matthias for enabling them to make the study and nowhere is there any mention of rendering magnifications, let alone the use of the space charge lens in an electron microscope. Therefore, the paper looks more like an effort to produce results that have a clearer relationship with the oscilloscope research that was officially pursued at the High Tension Lab.

Even if the paper had no immediate relevance to the development of an electron microscope, however, it was undisputably von Borries' and Ruska's first joint paper. Many more were to follow, always with von Borries as first author and Ruska as second. In 1932 this was a logical order, as von Borries was formally Ruska's superior then. Hedwig von Borries suggests that it was simply the alphabetical order that they applied.

7.2 Adolf Matthias

The intense activity of the two men coincided with Ruska's formal request to the head of the High Tension Laboratory to be allowed to build an electron microscope. The letter that Ruska wrote to Matthias on 13 April 1932, carried as header:

> *"Message about the possibilities for physical experiments and technical tests of materials, offered by an electron tube that is designed as an electron microscope" (von Borries, 1991).*

In the letter, Ruska expressed his wish to image objects that either emit electrons, are permeable to electrons or reflect electrons, and to study the geometry of the electron trajectories and the interactions between object and beam. Meanwhile, his description of the device that he would like to construct reveals a type of apparatus that can clearly be recognised as an electron microscope in its contemporary meaning:

> *"For a fruitful examination of the described possibilities an apparatus is needed, which is specifically suited for* electron microscopy, *that is, an electron tube which allows for electron optical magnifications in one or more stages. So far, concentrating coils have proven to be the best lenses, which have to be suitably designed here as objective lenses and projector lenses. Furthermore, it is necessary to add ocular micrometers which are adjustable in vacuum, and viewing screens at the location of the image and intermediate image" (Ruska, 1979). [Emphasis by JvG.]*

The intermediate viewing screen refers to the patents which had been filed four weeks before by von Borries and Ruska. Despite this, making enlargements is not mentioned as a research goal as such in Ruska's letter. Historian Qing Lin (1995) has interpreted this as evidence that the magnifying potential of an electron microscope was not Ruska's primary interest. To me, a more likely explanation is that the letter reflects a need to somehow fit the research proposal into the overall research plan of the High Tension Laboratory.

The appearance in the letter of the words "electron microscopy" deserves special attention, since it is the first time ever that we know of that these two words were written down, although not in an official publication. The use of the two words unambiguously marks the beginning of what was apparently experienced as a nascent discipline by its author. According to Hedwig von Borries, however, the author had not acted on his own here, suggesting that the "electron microscope report" as Ruska's letter is called by her, was written by von Borries and Ruska together.[29] Her evidence is not very convincing, and could just as well indicate that Ruska had simply discussed his text with von Borries—something that would have been very understandable and appropriate when considering their intense cooperation. Mrs von Borries' statement that her brother has never mentioned any involvement of von Borries with regard to this letter, is certainly correct.[30]

The same letter to Matthias also provides an estimation of the total costs of building the apparatus. Altogether 2255 Reichsmarks were needed,

[29] Von Borries (1991). The relevant paragraph contains an error, saying that Ruska's report was dated 13 March instead of 13 April.

[30] The only place where Ruska has discussed the letter himself is in Ruska (1979, 1980).

including the costs of photographic material. This amount is close to what someone had to pay for a middle class sedan at the time. Matthias approved of the plan and arranged for the funding as Ruska told us plainly in 1985:

> *"I had written a letter to Matthias to say that I wanted to build this new apparatus, and then he got the necessary money granted by the* Gesellschaft von Freunden der Technische Hochschule*" (van Gorkom & de Haas, 1985).*

In retrospect, the intriguing question is of course why Matthias felt inclined to facilitate the whole enterprise. After all, as emphasised before, the lab's major focus was the oscilloscope and such an instrument does not need to be able to render large magnifications. Matthias' former assistant Grünewald stressed in 1976 that it was the primary task of Matthias to come up with a faster oscilloscope, but he adds:

> *"As director of the High Tension Lab and as a research professor, however, he had an equal interest to look sideways into new territory, while working on cathode-ray oscilloscopes" (Grünewald, 1976).*

Likewise, Freundlich wrote in 1994:

> *"Though the electron microscope does not fit into the original task set for the research team, Matthias was willing to support the development" (Freundlich, 1994).*

During the 1985 interview, when talking about the Knoll group in general, we asked Ruska about Matthias and the electron optical work:

Q "What was the role of Matthias?"

A "Oh well, Matthias did not bother at all about such details. He left it to Knoll."

Q "Did he visit the lab on a daily basis?"

A "Matthias hardly ever visited the place where Knoll's team was installed. We hardly saw him in that whole period."

Q "He spoke to Knoll?"

A "Sometimes he spoke to Knoll."

Q "And not to you?"

A "Hardly ever to us."

Q "Also later, when you were doing your PhD?"

A "Little, still. He was very hard to reach. I have many letters in which Herr von Borries makes bitter complaints that [Matthias] does not read through his PhD thesis and so on. So, he did not attend to us very much. He had many other occupations, had little time for us. But he neither intervened in our projects, nor wanted to be a co-author."

Q "That is the positive side of it?"

A "That was the positive side of it, indeed. I am very grateful towards him that I was allowed to work there, that nobody bothered me, and was allowed to do my own thing—that was very nice."

It appears that we have to conclude that Matthias simply wished to respect and even to support the scientific curiosity of his young students without feeling inclined to spend too much time on what they were actually trying to do. These were perfect conditions for the lone wolf that Ruska soon turned out to be.

7.3 Max Knoll

Although Bodo von Borries had taken over as personal assistant to Matthias on 1 April 1932, business with Max Knoll had not finished yet. On 16 June 1932 Knoll and Ruska submitted their third joint paper, which was their second paper on the electron microscope. It appeared in the *Zeitschrift für Physik* and this time it carried the very straightforward title *Das Elektronenmikroskop*. Right from the beginning, they make it clear that they mean to discuss an apparatus that is intended to produce electron optical magnifications and start by giving a rather loose definition of the device, stating that:

> *"With electron microscope we refer to an electron optical arrangement that serves to examine emitting or irradiated objects by means of magnified imaging, while at least the first stage of the imaging is carried out by electron rays."*

The words "electron microscope" are accompanied by a footnote that refers to Knoll's first presentation of electron optical results in the Cranz Colloquium a year before, which is clearly an attempt to lay a priority claim. In the first half of the article, the two authors look into the theory of all sorts of ways to produce electron optical images. Just as they had done in their previous article, they stress that higher acceleration voltages will improve the quality of the imaging, since in an imperfect vacuum this will reduce the scattering of electrons by residual gas atoms. As we saw before, this remark bears relevance to the opinion of Ernst Brüche at the AEG that high voltages are troublesome. Subsequently, the authors mention for the first time the need to pay attention to external magnetic fields that may influence the performance of an electron microscope, like electromagnetic stray fields of nearby electrical equipment or the earth's magnetic field.

These remarks are concluded by doubtless the most historical section of this paper, titled "Definition and limit of the resolution." Knoll and Ruska had finally become aware of the limit to the resolution of electron rays as set by their wavelength, which follows from de Broglie's equation $\lambda = h/mv$.

Applying a modified version of this equation and inserting the outcome in Abbe's equation, they made estimates of the theoretical resolution for acceleration voltages of 1500 V and 75,000 V, which gives values of 15 and 2.2 Å respectively, based on a numerical aperture of 0.02. These estimates were going to suffice for electron microscopy for at least the next forty years. Meanwhile, these calculations had also provided a second reason for applying high voltages in electron microscopes. The estimates were followed by just as historic words:

> *"The theoretical* limit *of the resolution of an electron microscope [emphasis by Knoll and Ruska] as given by the wave nature of electrons, is therefore at least two to three magnitudes larger than that of a common microscope and is of the order of atomic distances. Whether this high resolution can be used to visualise structures of the same magnitude, cannot be determined at the current state of research and remains subject to further development of the methodology, which apart from a closer study of imaging errors, will also have to comprise an increase of the intensity of the electron source. Research in this direction is being undertaken" (Knoll & Ruska, 1932a).*

In the second part of the article Knoll and Ruska treated their results, although in reality most of this part was still rather theoretical. One very tangible result was the comparison of a light optical image of a molybdenum mesh wire to an electron optical one of the same mesh wire. They concluded that the electron optical image might even be better. A second, though not experimental feat is the first presentation to the outside world of the polepiece lens. The lens had not been tested yet as the authors mention themselves, and as such this new lens is a rather confusing detail—at least from a historical perspective. After all, the polepiece lens was patented three months before by von Borries and Ruska, and now it appeared in an article by Knoll and Ruska, at the moment that von Borries happened to be Ruska's tutor, while Knoll had been Ruska's tutor at the moment the patents were filed.

In the third part of the article possible applications of the electron microscope were discussed. This was a systematic attempt to cover anything one could possibly think of. First of all they mention the idea of sending electrons through an object, the method which we call transmission electron microscopy nowadays. Practical applications, they thought, were the study of lattices, very small apertures, and differences in density and granules in thin layers. Another method, they thought, would be the study of surfaces by looking at the reflection of electrons. Studying self-emission was the next possibility, in which case the radiation source becomes the object itself; the imaging of emitting cathodes is the best example here.

They also describe one specific case of self-emission and that is the photoelectric effect, where the intense irradiation of a metal surface with light will trigger the release of electrons from that surface if an electric field is present. Furthermore they suggested that the self-emission of any other suitable substance could be studied by applying it as a thin coating to the cathode. A next application of the electron microscope could be the imaging of space charge—i.e. electron clouds—they believed. Finally they ended with the idea of combining an electron optical magnification with a light optical stage, adding quite rightly the prerequisite that the resolution of the fluorescent screen should be good enough in that case.

The last and fourth part of the article was dedicated to the *Ionenmikroskop* (ion microscope) as opposed to the *Elektronenmikroskop*. Here they describe a microscope that works with positive particles instead of negative ones, and provide some calculus for the use of alpha particles—that is, helium atoms without their two electrons—and argon ions. It is rather surprising that the authors do not base this new concept on de Broglie, but instead refer to the wave mechanics of Hamilton—a theory that already existed for nearly a hundred years at the time. Evidently, these kinds of surveys were intended to claim priority to some extent. Nevertheless, as we have already seen, the authors were too late as far as the combination with light optics was concerned or the use of positive particles. As mentioned in the previous Part, both ideas were described by Léo Szilárd in the patent that he filed a year before. And Rüdenberg too had already tried to patent the idea of adding a light optical stage on behalf of Siemens three months before.[31] The possibility of using positive particles was claimed by Siemens as well,[32] but in this case Knoll and Ruska were two days earlier. Finally, the idea of making use of the photoelectric effect had already been patented, as we will see in the following subsection.

Though the Knoll and Ruska article provides the very first estimate of the resolution of an electron microscope, it was not the first time ever that a reference had been made in print to de Broglie in connection to electron microscopes. The news about de Broglie's thesis had reached Knoll—most likely in early 1932—through Fritz Houtermans (Landrock, 2003), a remarkable physicist of Dutch–Austrian descent who in 1927 completed his

[31] Reinhold Rüdenberg, Einrichtung zum vergrößerten Abbilden von Gegenständen durch Elektronenstrahlen, *German patent 916 839*, priority date of 31 March 1932, published 19 August 1954.

[32] Richard Swinne, Einrichtung zum Abbilden von Gegenständen, *German patent 915 843*, priority date of 18 June 1932, published 29 July 1954.

PhD in Göttingen magna cum laude as a student of James Franck. This was two years after Franck received the 1925 Nobel Prize in Physics together with Gustav Hertz "for their discovery of the laws governing the impact of an electron upon an atom." From 1928 until 1932 Houtermans worked as a personal assistant to Gustav Hertz, who was director of the Physics Institute at the Technische Hochschule of Berlin. It is not clear how Knoll got in touch with Houtermans, but since both of them worked with electrons, and both were at the Technische Hochschule, it is not at all surprising. Houtermans' primary focus was nuclear physics, but somehow the novel concept of the electron microscope must have caught his imagination, and this resulted in a joint study with Max Knoll and Werner Schulze, a PhD-student of Hertz.

The results of their study were presented for the first time on 14 February 1932 in a lecture, given in Braunschweig at a regional meeting of the German Physical Society by Fritz Houtermans. This was in about the same week as the appearance of the long Knoll and Ruska article in the *Annalen der Physik*, in which Ruska's electron microscope was described for the very first time (Knoll & Ruska, 1932b). Houtermans' talk appeared as an abstract in the *Verhandlungen der deutschen physikalischen Gesellschaft* and was called "About the geometric optical imaging of hot cathodes by electron beams with the help of magnetic fields (electron microscope)" (Knoll, Houtermans, & Schulze, 1932a). As we see, the term "electron microscope" was put between brackets, just as in the Knoll and Ruska paper in the *Annalen*, which illustrates once more how novel the concept still was in early 1932.

In his lecture, Houtermans directed his focus primarily to the possibilities of imaging the cathode's self-emission, as it is called in the list of applications that Knoll and Ruska were going to publish four months later. Furthermore, he also discussed the imaging of electron clouds—another application that was to be listed in the Knoll and Ruska paper. From a historical point of view, however, the abstract of the lecture is especially important for the fact that it is the first time ever that a remark appeared in print about the de Broglie wavelength in relation to electron microscopy:

> *"The resolving power of the electron microscope is—when applying Abbe's formula —in principle only restricted by the de Broglie wavelength of the used electrons, and can therefore theoretically be increased far beyond the limits given for the common microscope."*

Four months later, the same results were also presented by Knoll himself at a meeting of the German Physical Society and the Society of Applied

Physics on 17 June 1932—exactly one day after the submission of his third article with Ruska. This can be concluded from a footnote in the comprehensive report that Knoll, Houtermans and Schulze submitted to the *Zeitschrift für Physik* on 9 August 1932 (Knoll, Houtermans, & Schulze, 1932b) This final publication by the three men was called "Study of the emission distribution on hot cathodes with the magnetic electron microscope". Just as in the third Knoll and Ruska article that was submitted two months before, the words "electron microscope" were used freely now. The article shows that Knoll, Houtermans and Schulze constructed an electron microscope that was quite similar to the prototype that Ruska had built the year before. The major difference was the use of a hot cathode, and a horizontal column instead of a vertical one. Apart from that, they too mounted two lenses at the outside of the tube—in other words, the wall of the tube ran through the centre of the lenses—and they had also fitted their lenses with an iron casing on the outside and inside with a small circular gap at the inside. The article discussed exclusively the hot cathode's self-emission, which was illustrated with no less than 20 images, and also included the observation of electron clouds. No word whatsoever was spent on either the electron microscope's resolution, or alternative applications of an electron microscope. The emission by the cathode had their sole interest, while this type of research was rapidly becoming fashionable, as we will see in the following section on AEG. It should also be added that cathode studies must at least have been the main focus of Knoll, since he was working in the television industry now.

This is where the story of these three men would have ended, if we go by what they chose to make public at the time. The archives, however, show that they undertook one more significant action during their period of cooperation. This was the joint application for a patent, which was filed in their names on 17 March 1932—a remarkable day, since it is the same day that the two patents of Von Borries and Ruska were filed. The patent of Knoll, Houtermans and Schulze was given the title *Elektronenmikroskop, bei dem Elektronen aussendende Substanzen in vergrößertem Maßstabe abgebildet werden* (Electron microscope, which produces enlarged images of substances that emit electrons) (Knoll, Houtermans, & Schulze, 1932c) and it claims very specifically the invention of a device to study substances which emit electrons in response to irradiation. A thin layer of a substance of interest is to be mounted on a cathode where it can be irradiated with electrons, light or X-rays. Electromagnetic coils turn the emitted electrons into an image. In Fig. 6 we see this illustrated for light, which is emitted by its

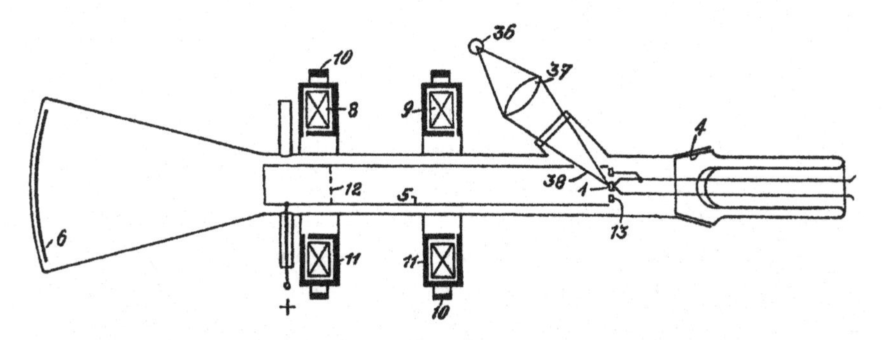

Figure 6 First design of a photoemission electron microscope. From Knoll et al. (1932c).

source indicated with the number 36, and focused on the substance of interest (1) by means of a lens (37). Basically, this patent describes the invention of what nowadays is known as the photoemission electron microscope (PEEM).

The German patent was followed by a US-patent application in June 1934.[33] Although the American patent derives its priority from the German original, it is markedly different. It carries the rather nondescript title "Electron microscope" and the emphasis lies primarily on self-emission in general, while photoelectric emission is only mentioned in passing. The patent contains a drawing, which is nearly identical with Fig. 6, but lacks the light source as well as its mount. As such, the drawing in the US patent is identical with the drawing that the inventors had published in the comprehensive article in the *Zeitschrift für Physik*. At the same time more emphasis is given to the use of reflected electrons, and there is the explicit reference to ions as alternatives to electrons, which is not present in the German patent.

Altogether, we start to get the image of a peculiar situation at the High Tension Laboratory in early 1932. On one hand von Borries and Ruska decided to draw up detailed plans for a transmission electron microscope, as we call it in retrospect. On the other hand, Knoll took refuge in the Hertz laboratory to satisfy his apparent interest in what became known as emission microscopy, and which resulted in a patent on the photoemission electron microscope, while he was still Matthias' assistant. The fact that von Borries and Ruska applied for their patents on the same day as Knoll, Houtermans and Schulze is certainly intriguing when taking the fear of

[33] Max Knoll, Fritz Houtermans and Werner Schulze, Electron microscope, *US Patent 2,131,536*, priority date of 23 June 1934, granted on 27 September 1938.

competitors into consideration, which Ruska had expressed in his letter to Matthias the day before. Apart from that, you may wonder why Knoll chose to join Houtermans and Schulze. Does it mean that there was not sufficient interest in self-emission at the High Tension Lab? Is this also a reason why he moved to Telefunken soon afterwards? Finally, our knowledge of Knoll's recent patent also changes the image that we have of the third joint article that he and Ruska completed in June. Not only does the article contain a hidden reference to the von Borries–Ruska patent on the polepiece lens, but the suggestion to make use of the photoelectric effect also turns out to be a hidden reference to a patent. Nowhere is there any mention of these patents, while at the same time Fritz Houtermans is explicitly credited in a footnote with another suggestion, that is, the idea of imaging foreign substances by mounting them on the cathode.

7.4 The de Broglie Wavelength

The exact moment at which Ruska was informed by Knoll about the 'de Broglie wavelength', as it became known initially, is very uncertain. In the speech that Ruska gave in 1970 upon receiving the Paul-Ehrlich- and Ludwig-Darmstaedter-Prize, he said that it happened in early 1932 (Ruska, 1970a). In the historical contribution that Ruska published in 1979, he states that he learned about it halfway through 1932 (Ruska, 1979). In our interview of 1985, he told Bart de Haas and me as well that it had been in 1932, but later on the date changed to June 1931. And a year later he asserted in his Nobel lecture that he had heard of it for the first time in the summer of 1931 (Ruska, 1987). The summer of 1931 is rather unlikely, however, as there is no reference whatsoever to an electron wavelength in the second joint article, which Knoll and Ruska submitted on 10 September 1931.[34] Nor is there any mention of de Broglie in the two patents or the article that von Borries and Ruska published in the spring of 1932. This is not a decisive argument, of course, but on the other hand we do see a sudden appearance of references to electron wavelengths from late spring 1932 onwards. Rüdenberg mentions them for the first time in his US patent on the electron microscope that he applied for on 27 May 1932,[35] Knoll and Ruska make their estimates of electron wavelengths in the paper that they

[34] Max Knoll and Ernst Ruska, Beitrag zur geometrischen Elektronenoptik (I and II), *Annalen der Physik* 404 (1932) 607–661.

[35] Reinhold Rüdenberg, Apparatus for producing images of objects, *U.S. Patent 2,058,914*, priority date of 30 May 1931, granted on 27 October 1936, p. 2, lines 6–10.

submitted on 16 June 1932. AEG's Ernst Brüche made his first reference to de Broglie during a lecture that he presented on 17 June 1932 as we will see in the next section, and on 18 June 1932 Richard Swinne filed a patent on behalf of Siemens which was primarily meant to establish the link between the electron wavelength and the superior resolution that could be obtained with it.[36] Altogether, it seems more likely that Ruska learned about de Broglie somewhere in April or May 1932.

Ruska also told us in the 1985 interview that Knoll broke the news to him during the short walk that took them from the Technische Hochschule to *Bahnhof Zoo*—the main railway station opposite Berlin's famous zoological garden. In 1970 he described the shock that this news had given to him:

> *"Even today I remember well how Knoll told me for the first time about this new wave property, as I was very upset about the fact that again there was a wave issue that would limit the resolution. I was only relieved when it became clear to me, when applying the de Broglie equation, that these waves were some five magnitudes shorter than light waves" (Ruska, 1970a, 1979, 1987).*

From a retrospective point of view, it is rather puzzling of course that it took so long for "Busch to meet de Broglie," or in other words, for the quantum physics of electrons to become integrated into the nascent field of geometrical electron optics. It lies far beyond the scope of this text to attempt to address the question how that could have happened, but the history of the electron microscope provides another interesting example of the gap between traditional physics and quantum physics in Germany during the 1930s. In 2005, Robert Perry published his memoir of Thomas F. Anderson, a physical chemist who became one of America's greatest pioneers of applied electron microscopy. Perry tells us about a young Tom Anderson:

> *"After receiving his B.S. degree in 1932, Tom spent a year in Kasmir Fajan's* Physikalisch-Chemisches Institut *in Munich. This laboratory was concerned with the determination of refractive indices of various substances to the highest possible degree of precision. (...) The interpretation of [Tom's] data was made according to old classical theories of refraction developed by Lorentz, rather than the newer quantum mechanics, which had not yet effectively penetrated the thinking of the Munich group. When Tom returned to Caltech, he was persuaded to give a seminar on this research before an audience that included Linus Pauling. About midway through the seminar, Pauling commandeered the blackboard and, much to Tom's chagrin, sketched out the quantum theory of mole refractions" (Perry, 2005).*

[36] Richard Swinne, Einrichtung zum Abbilden von Gegenständen, *German patent 915 843*, priority date of 18 June 1932, published 29 July 1954.

7.5 Interlude

After the submission of the third Knoll and Ruska article in June 1932, Knoll's role had finally ended, and Ruska was left to continue business with von Borries alone. Going by the information that Heinrich Grünewald and Hedwig von Borries derived from Bodo von Borries' calendar notes, the two were spending many nights and weekends on their joint projects, but whatever they were working on, these efforts did not immediately result in tangible outcomes for the remainder of the year. From Mrs von Borries' memoir it can be learned that they had been working hard on the *Ruska Arbeit*, as Bodo wrote to his parents on 17 September 1932. The German word *Arbeit* can mean anything from paper or study to assignment or job—just like its English counterpart "work"—so one can only guess what he meant to say. Most likely, it referred to the construction design of the polepiece lens, which was going to be the subject of Ruska's doctoral thesis. On 1 December 1932, they did submit a short note, though, to the *Archiv für Elektrotechnik* (von Borries & Ruska, 1933a). In it they challenged Walter Rogowski, who had published an article with the title "On the explanation of the gas concentrated electron beam". Rogowski, professor at the Technische Hochschule of Aachen, was one of Germany's well known developers of cathode-ray tubes and additional devices. Just as von Borries and Ruska had done in the first joint article that they had submitted earlier that year, Rogowski and his colleague Graupner explained the dynamics of the *Fadenstrahlen* (thread beams) and suggested a practical application which was very similar to von Borries' and Ruska's space charge lens. Rogowski and Graupner, however, made no reference whatsoever to von Borries and Ruska's earlier work, despite the fact that their article had been submitted by the end of September, while the von Borries and Ruska paper had appeared in July. Understandably, von Borries and Ruska felt challenged to set this record straight. At the same time, we should remind ourselves that this work on space charge did not bear any immediate relevance to the development of an electron microscope.

Finally, the year 1933 arrived and everything was going to be different again. According to Hedwig von Borries' memoir, Bodo's father decided to retire on health grounds, and this would have been the reason for Bodo to apply for a properly paid job in order to relieve his parents financially. He soon found a position as a graduate engineer with the *Rheinisch-Westfälische Elektrizitätswerke* in Essen, which meant that he was to give up his position at the TH by 1 March 1933, as he informed Adolf Matthias on 2 January.[37] On the same day, von Borries drew up a signed agreement together with Ruska as well as Freundlich. Apparently, a need was felt to secure future credits now von Borries was leaving.

The agreement describes five articles that the three of them wished to publish. Von Borries and Freundlich intended to publish together about an "electron beam relay" and "linear-time-base deflection with the cathode-ray oscilloscope", von Borries was to publish a paper together with Ruska about "innovations of the electron microscope design", Ruska himself would write a paper about "very large magnifications with the magnetic electron microscope" and finally a paper was foreseen on a new way of "measuring currents with a cathode-ray oscilloscope", which was to be written by Ruska, von Borries and Matthias (Grünewald, 1976; von Borries, 1991; Freundlich, 1994; Lin, 1995). So in fact, only two of the five articles were to be dedicated to electron microscopy, in both cases involving Ruska, while the article on very large magnifications that Ruska was going to write on his own, was certainly the most outspoken one with regard to microscopy.

7.6 The Rise of Nazism

In 1933 as well, politics already started to play a serious role, as has been indicated before in Section 3. Carlheinrich Wolpers, who became involved with German electron microscopy in 1938, is the only eye-witness who ever wrote about the political background against which the instrument's development took place. In the recollections which he published in 1991, he paid attention to the political unrest that had been growing for some time already:

> *"In the depressing time between 1929 and 1932, the number of unemployed in humiliated Germany had risen from 2 to 6 million, the rightist "Harzburger Front" and the leftist "Iron Front" were continually engaged in bad bloody fights, seething with hatred, and the worldwide economic crisis led to loss of wages and to deductions from salaries (. . .); despite these conditions Knoll and Ruska finished in the spring of*

[37] Hedwig von Borries, Bodo von Borries, *Advances in Electronics and Electron Physics* 81 (1991) 127–176, p. 137. According to Martin Freundlich the actual date on which von Borries left the lab was 18 February 1933: Martin M. Freundlich, The History of the Development of the First High-Resolution Electron Microscope, *Microscopy Society of America Bulletin* 24 (1994) 405–415, p. 411. Grünewald writes that von Borries started in Essen on 21 April: Heinrich Grünewald, Zur Entstehungsgeschichte des Elektronenmikroskops mit elektromagnetischen Linsen, *Technikgeschichte* 43 (1976) 213–222, p. 217.

1931 the construction of the first two-stage transmission electron microscope with magnetic lenses" (Wolpers, 1991).

The leftists were members of the communist party, and the Harzburger Front was a broad coalition in which Hitler's SA (*Sturm Abteilung*, or Storm Division) took part as well. A quote from the 1985 interview with Ruska confirms this image of the events in the streets of Berlin in those days:

> *"... So I have known the old Berlin, East-Berlin actually, without any Hitler business, during the Weimar Republic. And then these complete street fights and rows with which Hitler slowly gained his influence—with his SA folks and so on. Then you often had these street fights, also close to where we lived at* Fehrbelliner Platz *(...), in which occasionally people got killed. I have seen all of that too."*

1933 is especially the year in which Hitler managed to take control of Germany. After the federal elections of 31 July 1932, Hitler's National Socialist German Workers Party (NSDAP) had increased from 18 to 37 percent of all voters. This made them the largest of more than ten parties in the *Reichstag* (parliament), but they still needed to form a coalition with other parties in order to obtain a parliamentary majority. The establishment of such a coalition failed, and so there were already new elections again on 6 November 1932, in which the Nazis actually lost 4 percent, going back to 33 percent. This still made the NSDAP the biggest party, but again it turned out to be impossible to form a party coalition that could rely on a majority in the Reichstag. Once more there were going to be elections, now on 5 March 1933. Then something happened that was very convenient for Hitler. In the evening of 27 February 1933 the Reichstag had been set on fire at many places, and the building had blazed like a bonfire. The arsonist, the Dutch bricklayer Marinus van der Lubbe, was quickly arrested. It was immediately decided that the arson was a communist act of terror, and the next day an emergency decree was issued, known as the *Reichstagsbrandverordnung* (Reichstag Fire Decree), which took away most constitutional rights. Six days after the fire the planned elections were held, giving Hitler 44 percent of the seats in parliament.

The author Christopher Isherwood, who lived in Berlin at the time, has provided us with a literary impression of the atmosphere afterwards:

> *"Early in March, after the elections it turned suddenly mild and warm. 'Hitler's weather,' said the porter's wife; and her son remarked jokingly that we ought to be grateful to van der Lubbe, because the burning of the Reichstag had melted the snow. (...) Uniformed Nazis strode hither and thither, with serious set faces, as though on weighty errands. The newspaper readers by the café turned their heads and smiled and seemed pleased. They were pleased because it would soon be summer, because Hitler had promised to protect the small tradesmen, because their*

newspapers told them that the good times were coming. They were suddenly proud of being blonde" (Isherwood, 1935).

Soon afterwards Hitler managed to convince enough of the smaller conservative parties to vote for the so-called *Ermächtigungsgesetz* (Enabling Act) that passed the legislative powers from the Reichstag to the government.[38] After this Act came into force on 24 March 1933, Hitler could draw up any law that he wished.[39] This date marks the official end of German democracy and the beginning of Hitler's Third Empire. On 10 May 1933 the notorious Nazi book burnings took place, one of them near the Reichstag, where students and professors with Nazi sympathies burned some 25,000 books that had been labelled "un-German." On 14 July 1933, finally, the *Gesetz gegen die Neubildung von Parteien* (Law against the foundation of new parties)[40] outlawed all political parties except the NSDAP. Carlheinrich Wolpers described how he perceived this episode:

"With some help from the bourgeoisie, taking advantage of the weakness of the 85-year-old president, in 1933 Hitler suddenly reached a powerful position (...) the dictatorial regime had a firm seat and interfered with growing pressure in the habits of life of everybody within the practically closed frontiers. Adaptation, retirement and distrust led us to adopt a pragmatic behaviour" (Wolpers, 1991).

A political development with immediate effect on the Technische Hochschule was the *Gesetz zur Wiederherstellung des Berufsbeamtentums* (Law for the Restoration of the Professional Civil Service), passed on 7 April 1933.[41] As a consequence of this law, anyone who was fifty percent Jewish or more was not allowed to be employed by public bodies. As mentioned in Section 3, one of the victims of this new law was Gustav Hertz, to whom Siemens offered the opportunity to continue to work after he was forced to leave the Technische Hochschule. But Freundlich too was going to be affected, although not immediately. As was mentioned above, he succeeded von Borries as assistant of Matthias from 1 March onwards. One of his tasks was, he wrote in 1994:

"(...) to supervise the printing of the doctoral theses of the five members of the research team. Though all five had been printed and were ready to be bound and published, only three appeared in Forschungsheft #3 *of the* Studiengesellschaft,

[38] Gesetz zur Behebung der Not von Volk und Reich, *Reichsgesetzblatt, Teil 1* (1933) 141.

[39] More on the Reichstag Fire Decree and the Enabling Act on the website of the German Federal Parliament, http://www.bundestag.de/htdocs_e/artandhistory/history/factsheets/enabling_act.pdf (accessed 27 October 2012).

[40] Gesetz gegen die Neubildung von Parteien, *Reichsgesetzblatt, Teil 1* (1933) 479.

[41] Gesetz zur Wiederherstellung des Berufsbeamtentums, *Reichsgesetzblatt, Teil 1* (1933) 175.

> *namely those of Knoblauch, von Borries and Ruska. Those of Lubszynski and myself were suppressed due to Nazi persecution."*[42]

And a little further he added:

> *"Professor Matthias informed me in May 1934, that as a Jew, he could no longer keep me in his employment. The cathode ray oscillograph that I had constructed was 95% completed. The finishing touches were left to Dipl.-Ing. Mochow. I left as the last member of the Knoll team. I emigrated to England at the beginning of September 1934, and live, since September 1936, in the USA."*

This not only illustrates the effects of politics on everyday research, but also the fact that Matthias took more than a year to follow up the law that forced him to send away Freundlich.

For Ernst Ruska Nazism was to become even more personal, as his own friend Bodo von Borries decided to join Hitler's NSDAP, which made him officially a Nazi. According to the NSDAP membership records, which can be found at the Bundesarchiv in Berlin, he became a member in Essen on 1 May 1933.[43] It is one of the better kept secrets of the early history of electron microscopy. As far as I know, no one has ever published anything about it, although an indirect reference to it was made in a historical recollection, written in 1985 by the Dutch pioneer Jan Bart Le Poole, who wrote that von Borries was wearing a swastika on his lapel during a visit to the Netherlands in 1942.[44] According to Ernst Ruska and his wife Irmela, von Borries had not joined the party out of idealism, but had regarded it as his duty, as he came from an aristocratic background. On 22 May 1905, Bodo von Borries had been born in Herford—a German town in the state of Nordrhein-Westfalen, which borders the Netherlands and Belgium. His father Franz von Borries was *Landrat* of the Herford district—a kind of governor. Since 1733, the von Borries family belonged to the German elite, after Johann Friederich Borries—the great-grandfather of Bodo's grandfather—was admitted to the nobility by Emperor Charles VI.[45] From then on the family had the right to add the aristocratic prefix "von" to their surname.

[42] Freundlich (1994). "Forschungsheft" means research journal. "Studiengesellschaft" refers to the Research Society for Very High Tension Facilities. Hans Gerhard Lubszynski became a well-known television engineer in Britain in later years.

[43] Membership number 3255931, Bundesarchiv, Berlin-Lichterfelde, BDC Reichskartei film C0102 and BDC Ortsgruppenkartei film C0004.

[44] Le Poole (1985). This anecdote has later been cited in Fournier (2009).

[45] http://www.familievonborries.de, accessed 27 October 2012; Ernst Heinrich Kneschke, *Neues allgemeines deutsches Adels-Lexicon*. Leipzig: Friedrich Voigt, 1859, Vol. 1, p. 582.

Evidently, serving your country is considered to be a noble trait, and for some reason this appears to have implied that one should support the Nazi regime, despite the well-known fact that Hitler did not like aristocrats at all. In the 1985 interview, Ernst and Irmela Ruska were eager to play down the significance of von Borries' membership. Mrs Ruska told us for example:

> *"That is this missionary zeal of the aristocrats again. They always wanted to be part of it when it comes to politics, they always wanted to take the lead by setting examples, to be a role model."*

A little later Ernst Ruska remarked:

> *"That is this frequently discussed problem between Germans and non-Germans after Hitler. There were so many people in the party, but these were not conscious murderers of Jews. This treatment of Jews seeped through only gradually [to the consciousness of the general public]."*

Irmela Ruska subsequently added:

> *"Herr von Borries was a 'good German,' you might say. He believed—so he had said—that 'if I do not join the war, than I have to serve my country in another way.' And then he told my husband, 'you don't join them because you are too lazy.' "*

At the same time, Mrs. Ruska was eager to stress on more than one occasion that her husband never followed his friend's 'good' example. A check in August 2013 of the NSDAP membership records in Berlin has shown that she was right.

Ernst Ruska on his part also wished to emphasize that von Borries had just been doing his duty, and he used their Jewish colleagues Freundlich and Lubszynski to illustrate this:

> *"All I want to say is that we had two Jews in our group with Knoll, and I have a letter from the one that went to America—and who has actively helped me against Rüdenberg—who writes to me that he is very grateful towards Herr von Borries for the way that he helped him with his emigration. Such things can go on inside a person at one and the same time; that one respects the laws of decency, or friendship, while despite of this one supports a political view of which nobody knew at the time in what incredible way it would turn against the Jews in the end. (...) Both Jews in our group emigrated in 1933 and all of us have helped them to get out safely. And this is what Freundlich has confirmed to me just a few years ago."*

7.7 Ernst Ruska—Left to Finish the Job

From 1 March 1933 on, Ruska was more or less on his own. Knoll had left 11 months before, and now von Borries was gone as well. The only person still around was Freundlich—who was even Ruska's superior now,

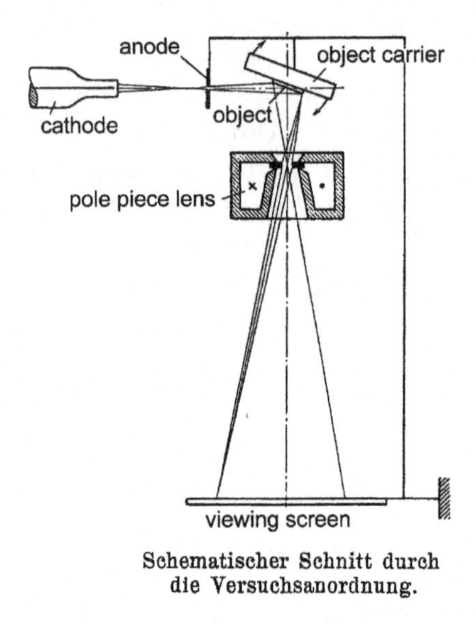

Figure 7 Schematic cross section of experimental arrangement. Based on Ruska (1933a, Fig. 1).

technically speaking—but he was not immediately involved with the development of the electron microscope. Ruska had come close to completing his PhD, which as said before, was dedicated to the construction design of the polepiece lens (cf. Fig. 7). It meant that he should start to look for a paid job, just like von Borries, as he was going to turn 27 that year and he could not expect his parents to continue to maintain him. He did not intend to quit, however, as it had not only been his intention to construct a polepiece lens, but also use it to build an improved electron microscope. Therefore, one of the first things that Ruska did after von Borries had left, was to apply for a second grant—this time with the *Notgemeinschaft der deutschen Wissenschaft* (Relief Association of German Science), the precursor of the present *Deutsche Forschungsgemeinschaft* (German Research Foundation). The requested grant was only supposed to supply Ruska with some income and was not needed for the instrument itself, since nothing new was needed that had not already been covered by the sum that had been supplied by the Society of Friends of the Technische Hochschule. Apparently Ruska's father was still quite supportive. To help his son get the money, Julius had even made use of his position to contact Max von Laue, who acted as an advisor to the Notgemeinschaft, as Ruska told us in 1985. It is impossible to tell whether this will have made any difference, but the request was granted on 22 June, and from 1 July 1933 onwards Ruska was going to receive 100 Reichsmarks per month for the remainder of the year.[46]

Qing Lin had access to the text of the application, which was submitted on 13 March 1933. Lin considers it to be the first document that contains a clearly defined goal to surpass the resolution limit as set by the light microscope. Indeed, the text of the application is far more outspoken than the application which Ruska submitted a year before, as it says:

> *"The studies should serve the further development of the electron microscope and its applications. Especially the realisation of the next two objectives is planned:*
> *1. Construction of an optimally adjusted electron optical device, with a magnification of at least 10,000 to image ultramicroscopical structures.*
> *2. Development of an experimental setup to obtain sharp electron images of transilluminated metal foils and illuminated metal surfaces. (. . .)" (Lin, 1995).*

Apart from the objective to surpass the resolution limit, this is also the first time that we see the adjective "ultramicroscopical" appear in writing.

Meanwhile, the departure of von Borries did not mean that the exchange of ideas between the two friends had stopped. Both Heinrich Grünewald and Hedwig von Borries emphasise that the joint sessions were replaced by a steady flow of letters. Mrs von Borries writes for instance:

> *"In some 170 pages of correspondence written by the end of 1933, a lively exchange of ideas went on as regards the development, construction and conducting of experiments that were to be carried out and had been jointly planned prior to von Borries' departure. Ruska reported regularly on all experiments, asking for von Borries' opinion when difficulties arose. He received suggestions and design diagrams in return" (von Borries, 1991; Grünewald, 1976).*

Soon after von Borries' departure, time had also come to submit the second joint paper by von Borries and Ruska.[47] This was done on 7 April 1933, which happened to be the same day on which the Law for the Restoration of the Professional Civil Service went into effect, which was going to terminate Freundlich's employment. The new article was called *Die Abbildung durchstrahlter Folien im Elektronenmikroskop* (The imaging of transilluminated foils in the electron microscope). It had already been announced in Ruska's letter to Professor Matthias in order to obtain his first

[46] Lin is very specific about the grant's end date of 30 June 1934 instead of 31 December 1933. Ruska himself, however, was not consistent at this point. In Ruska (1979) he mentioned that is was a grant for a year, in the 1985 interview he said that is was half a year and also in his Nobel lecture (1986) he spoke of half a year. See Lin (1995), p. 61.

[47] The third if you wish to include their response to the paper by Rogowski and Graupner.

grant, and as we just saw, was also mentioned in the application for the second. The article showed some electron optical transmission images, made in one stage, of sheets of aluminium with a thickness of 1 and 6 microns. The results confirmed the idea that electron rays can be used to render sharp images that reveal all sorts of detail. It is noteworthy that the article already contained thoughts on electron optical image formation, including some references to scattering, although the authors had not yet realised its significance, as will become clear later on. It should also be noticed that the polepiece lens was employed for the very first time in this study, which illustrates that Ruska's PhD was coming to an end.

Within another four weeks, Ruska published an article on his own, also a study that had been announced in his two applications for grants. Already the title made clear that he was now consciously taken the step from 'apparatus' to 'instrument'—from microscope to microscopy. The paper was called *Die elektronenmikroskopische Abbildung elektronenbestrahlter Oberflächen* (The electron microscopical imaging of surfaces irradiated with electrons), submitted to the *Zeitschrift für Physik* on 2 May 1933 (Ruska, 1933a). It is a rather semantic exercise, but the use of the adjective 'microscopical' automatically implies that the concept of 'electron microscopy' had now become fully accepted by a respectable journal as well. This is confirmed in the first sentence of the editor's summary, with which the paper begins:

> *"The construction is described of a magnetic electron microscope for viewing surfaces with reflected electron radiation, and with two images it is also shown that* also this kind of electron microscopy *[emphasis by JvG] will produce images of such surfaces."*

It is unlikely that Ruska knew that his former teacher Knoll had meanwhile patented the idea of making images with reflecting electrons, together with Houtermans and Schulze, as there is no reference to it in Ruska's paper. In it, an arrangement was demonstrated as shown in Fig. 4. The object was irradiated sideways and the reflected electrons that bounced off the object would travel through the lens to the viewing screen. For this purpose, Ruska had applied the same new polepiece lens that had been presented by him and von Borries in their previous paper. With this arrangement, he made images that even showed the 'electron shadow' of an extruding detail on a metal sheet, and produced the mirrored image of the number 71 that he had scratched into metal. In a footnote he expressed the intention to place the photographic film inside the vacuum of the microscope. In that case he would not depend on the rather low quality of the luminescent viewing screen, and besides that, he added, it would also be necessary for the planned recording of magnifications of 10,000 and more. Regarding the fact that no one in the world was able to make such large magnifications yet, this was a rather bold remark, but it was perfectly in line with the text of his grant application.

On 12 May 1933, so only ten days later, Ruska submitted another article on his own to the same journal. This one was titled *Zur Fokussierbarkeit von Kathodenstrahlbündeln grosser Ausgangsquerschnitte* (On the ability to focus bundles of cathode rays with large initial cross-sections) (Ruska, 1933b). It had neither been announced in one of the applications for a grant, nor was it part of the signed agreement with von Borries and Freundlich. The paper was certainly not a treaty on magnifications or electron microscopes. On the contrary, it was a theoretical exercise about concentrating the emission of a large electron source in one point. For cathode-ray oscilloscopes this is certainly a relevant issue, as explained at the beginning of Section 1. One needs a very bright dot on the oscilloscope's screen in order to be able to register extremely fast events. A way to accomplish this, is to use a larger electron source as it will emit more electrons. Nevertheless, the article could be interpreted as having indirect relevance to electron microscopy, since you also need a very intense beam when you wish to make very large magnifications. In that case the beam will not be concentrated on the screen, but on the object. Already a year before Ruska and Knoll had announced in their third joint paper that a study to increase the intensity of the electron source was being undertaken.[48] And on 1 December 1933, von Borries and Ruska would file a patent on a stronger cathode, as will be discussed in more detail in the next pages.

Ruska's last article as a PhD-student was finally one of the five papers that were mentioned in the signed agreement between von Borries, Ruska and Freundlich. It was about *Eine neue Form des Strom-Messsystems am Kathodenstrahloszillographen* (A new design of the [electrical] current measurement system for the cathode ray oscilloscope) which Ruska published together with von Borries and Matthias. Once more it was published in the *Zeitschrift für Physik*, this time submitted on 19 July 1933 (Matthias, von Borries, & Ruska, 1933). It dealt with a new type of deflecting coil to be used in a cathode-ray oscilloscope, which was designed to measure the intensity of electrical currents far more accurately than had been possible so far. To put it in the words of engineers, they had designed a more sensitive cathode-ray

[48] Max Knoll and Ernst Ruska, Das Elektronenmikroskop, *Zeitschrift für Physik* 78 (1932) 318–339, p. 329.

ammeter. As already observed before, this paper too had little to do with either the polepiece lens or electron microscope. With reference to the political circumstances it is noteworthy however that the article ends with,

> *"We owe many thanks to Mr Dipl.-Ing. M. Freundlich for his valuable help in recording the oscillograms."*

Since Freundlich had to be dismissed soon and as there were virtually no prospects for him in Germany, nobody would have remembered him if they had not mentioned him.

Ruska submitted his PhD-thesis on the polepiece lens on 31 August 1933. From that moment on, he was finally able to dedicate all of his time to the construction of the new microscope that would produce magnifications of at least 10,000 times. However, the need to look for a properly paid job was also becoming urgent now, as his allowance from the second grant was not going to last that long. Despite the infamous economic depression of those days, quite soon Ruska managed to find a reasonable position with *Fernseh AG*, a commercial manufacturer of televisions. It meant that at least he would stay in touch with electron optics to a certain degree, since basically the twentieth century television displays were also cathode-ray tubes. Ruska was expected to start in his new job on 1 December 1933.

This development shows that the results so far obtained by Ruska together with either Knoll or von Borries, had not been met with overwhelming enthusiasm by the outside world. If that had been the case, Ruska and von Borries would not have been forced to find employment in industry. As the following events make clear, at the same time Ruska cannot be blamed for lack of determination to pursue the development of the magnetic transmission electron microscope. On 23 November 1933—just a week before his research would end—Ruska submitted a short paper to *Forschungen und Fortschritte*, which was published on 1 January 1934 (Ruska, 1934c). The short article, called *Das Elektronenmikroskop als Übermikroskop* (The electron microscope as supermicroscope), appeared in a magazine that described itself as "Bulletin of German Science and Engineering", but was soon to develop into a Nazi propaganda tool that also appeared in English and Spanish translation. As the bulletin addressed a rather diverse audience, Ruska presented his paper as a general introduction to a new subject. However, it also appears to have been a deliberate choice to secure priority, as it contained major results of the last months which had not been published yet and which were only to appear in a later, far more comprehensive report. Ruska's short preliminary paper, covering less than a folio-sized page, is remarkable in many ways. First of all the title carries the word *Übermikroskop*, which was the first time that it appeared in print in such a prominent way. The word was not Ruska's own invention, however. Half a year before, AEG's Ernst Brüche had been the first to mint the word in his company's year book.[49] Nevertheless, the title as such was completely justified, as it was indeed Ruska's intention to tell the world that a microscope had been developed—*by him*—that was able to surpass the resolution limit of the light microscope. The use of an Übermikroskop, he explained, would be the study of "all sorts of objects, e.g. metal foils, very fine fibres and organic objects that are subjects of interest to physicians and biologists." He illustrated his plea with an 8400-fold magnification of a piece of aluminium foil, and added that 20,000 to 30,000-fold magnifications already belonged to the possibilities. Finally, the essay concluded with a strong appeal to the intended users of the instrument:

> *"With regard to the great fundamental significance of electron microscopes to Übermikroskopie it would be very desirable that—as well as physicists—physicians and biologists should participate in the necessary developments."*

Altogether, this paper provides the first hard evidence that it was Ruska's goal to construct an electron microscope which would not only render very high magnifications, but would also be instrumental in biology and medicine.

Just a week after Ruska had submitted his paper, von Borries and Ruska filed their third German patent as well. It describes the invention of an "Arrangement to generate a parallel or conical cathode-ray beam of high intensity," which carries the priority date of 1 December 1933.[50] The idea was to construct a trumpet-shaped cathode which would be mirrored by an anode of the same shape. The invention would be especially relevant to "cathode-ray oscilloscopes, television tubes and the electron microscope," they state in the very first sentence. As such, the patent could be considered to be a spin-off of the paper on large cathodes that Ruska had submitted half a year before.

As he had wanted to do some last experiments with his new apparatus, Ruska was already in his second week at his new job, when he finally submitted the comprehensive report on an improved electron microscope on 12 December 1933 (van Gorkom & de Haas, 1985). This

[49] Ernst Brüche, Geometrische Elektronenoptik, in: *Jahrbuch des Forschungs-Instituts der Allgemeinen Elektricitäts-Gesellschaft* 3 (1931/32) 111–124, published May or June 1933.

[50] Bodo von Borries and Ernst Ruska, Anordnung zur Erzeugung eines parallelen oder kegelförmigen Kathodenstrahlen großer Dichte, *German Patent 692335*, filed on 1 December 1933 and granted on 23 May 1940.

article also appeared in the *Zeitschrift für Physik*, carrying the title *Über Fortschritte im Bau und in der Leistung des magnetischen Elektronenmikroskops* (On advances in the construction and performance of the magnetic electron microscope) (Ruska, 1934a). The article reads like a do-it-yourself manual on building an electron microscope, providing detailed technical directions, followed by ideas for further improvements, and examples of what so far had been achieved with it. As such, this article appears not to be completely void of promotional intentions, just like the Übermikroskop paper that had been submitted to *Forschungen und Fortschritte* three weeks before.

Ruska begins the article by asserting that the imaging of cathode surfaces is not the only possible application of an electron microscope:

> *"The second area, which will be even more fundamental to our knowledge, is the visualisation of hitherto inaccessibly small particles and objects, which is in principle possible because of the extremely small wavelengths that can be assigned to electrons according to de Broglie."*

Here he was emphasising again the difference between his work and that of others. As I already remarked in the subsection on Max Knoll, it will become more clear in the next section that electron microscopy in the sense of (self-)emission microscopy had become very fashionable by then.

The microscope design that was presented in this paper, was an immense step forward in comparison to the contraption that had been constructed in spring 1931, which was little more than a cathode-ray tube fitted with two electromagnetic coils. In contrast to that device, the world's first *supermicroscope* already contained most of the basic features that were to become classic.

First of all, there was the polepiece lens, which was finally described in detail now; its concept had been patented one and a half years before. As can be seen in Fig. 8, its early design actually consisted of two separate coils on top of each other, each with its own circular gap at the inside of the iron casing. The two magnetic fields of the coils could be combined into a big one. As a result the centre of the combined field could be moved up and down by regulating the separate currents in the two coils. This would take away the need to move the object up and down, as one does in a light microscope. In the central passage of the "double objective"—as it was called—a screw thread had been tapped. This made it easier to position and replace the polepieces that further narrow the magnetic field. The thread could also be used to screw in a small piece of tube to close off one of the

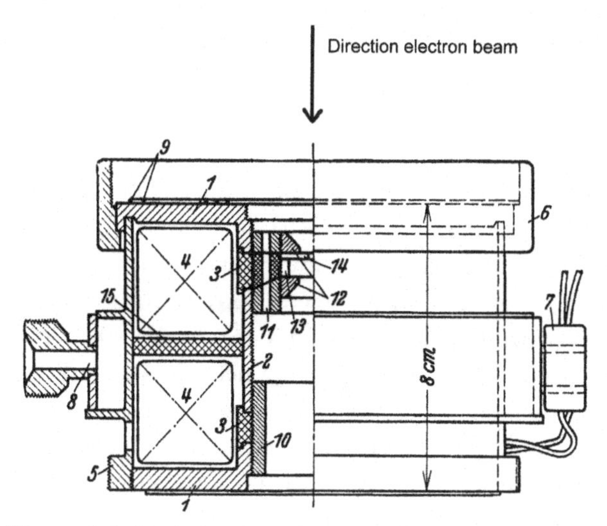

Fig. 6. Schnitt durch die Objektivspule (Doppelobjektiv).

Figure 8 Cross-section of the objective coil (double objective). Number 1 is the iron casing. Numbers 4 are the coils. Number 3 indicates the circular gaps. Number 12 points at the pole pieces. Number 10 is a piece of iron tube that can be used alternatively to seal off a gap and number 8 points at the water cooling. Image from Ruska (1934a).

two circular gaps. These kinds of technical practicalities were a reflection of the desire that Ruska and von Borries previously had had in mind to design an 'electron microscope construction kit', as it had been nicknamed in their correspondence.

The microscope had a total of three lenses. Apart from the objective lens, there was a condenser lens to concentrate the electron beam on the object, and a projector lens for the second magnification stage. The author already mentioned the option to use an extra lens between objective and projector lens, but this would only be needed "in very extreme cases," he believed (Ruska, 1934a). As a consequence of the new design, the lenses were located within the vacuum and all connecting parts of the tube were made of brass, in contrast to the contemporary experimental cathode-ray tubes, which were usually made of glass. The intermediate screen on top of the projector lens, which had been the subject of von Borries' and Ruska's other 1932 patent, had been included as well. Furthermore the instrument was fitted with a first primitive object chamber. The chamber contained a revolving object table with eight placeholders which made it possible to mount eight objects at once. The revolver could be operated from outside the vacuum in order to move one object after the other into the electron beam. The set of eight objects could only be replaced by taking the instrument apart, as there was no airlock yet. Finally, there was a water-cooling system to cool the electron gun, vacuum pump and electron lenses.

It had been the intention to make more adjustments, but there had simply not been the time or the money to do it. Instead, Ruska could only describe the desired changes. A major point of improvement would be a more stable electron gun. Small fluctuations of voltage and current diminished the quality of the electron images greatly. Also the need for an airlock was mentioned. A third point was the need for internal photography, as Ruska had pointed out half a year before as well (Ruska, 1933a) Photographs were still made from the outside, recording the images on the viewing screen. However, it would be far better if the electrons interacted directly with the photographic emulsion itself.

Probably one of Ruska's most important remarks was a theoretical insight. When making images of aluminium foils, it had finally dawned on him that electron images can be formed as a result of so-called elastic scattering instead of adsorption (Ruska, 1934a; van Gorkom & de Haas, 1985). The difference between the two processes is immense. In the case of absorption electrons become stuck in the foil, which will eventually result in evaporation of the aluminium, depending on the intensity of the beam. In the case of elastic scattering the electrons fly through the foil without any significant energy transfer from electrons to the aluminium atoms, although their interaction will cause a change of direction of the electrons. Already since the *Diplomaufgabe* that Ruska had pursued in 1930, aluminium foils had been a recurrent theme. At that occasion he had concluded that a very thin aluminium foil will scatter electrons in the same way frosted glass does. At the same time, Bodo von Borries had started his PhD-research on the outside photography of oscillograms for which he used Lenard windows. The whole principle of these windows is based on the application of a metal foil that is thin enough to let electrons pass through. The second joint paper by von Borries and Ruska in April 1933 also dealt with thin foils, so taking this all into account, it is not such a surprise that somewhere in 1933 Ruska began to realise that elastic scattering could explain image formation, implying that an object would not necessarily be incinerated by the electron beam. As he explained during the 1985 interview, all he actually had to do was to look up the percentage of absorption for certain foil thicknesses in Philipp Lenard's book *Quantitatives über Kathodenstrahlen aller*

Geschwindigkeiten (Lenard, 1918) in order to conclude that image formation had to be due to elastic scattering, when foils are thin enough.[51]

Another important result in the December 1933 article was the demonstration of the high resolving power of the new instrument. Ruska showed this by using a very fine cotton fibre. Ironically, such a fibre did exactly what any bystander would have predicted: they would simply char and evaporate, as they are far too thick for electron microscopy. However, in the very fast process of evaporation they would grow thinner and thinner, in that way giving a good idea of the smallest detail that could be distinguished. Ruska published the image of a charred fibre that was 0.3 micron wide at its thinnest spot. This is approximately the best a light microscope can show. Subsequently he pointed out that specific details in the image were even six times smaller, which actually meant that he had managed to surpass the magic limit to the resolution of the microscope. On the following pages, he presented several 12,000 fold magnifications of a charred fibre and grainy gold foil. These images will certainly have raised eyebrows with any critical reader at the time, as basically it must have been impossible to think anything of it, not having a clue of what you were actually looking at. This, however, did not deter Ruska from making a final bold suggestion in a last footnote:

> *"For a better visualisation of such objects—one might think of nerve fibrils with their very fine structures—one would have to develop 'colouring'-methods, for example impregnation with metallic salts (silvering), depending on the circumstances of the case; in the same way as is usual in the common histological practice concerning normal microscopy" (Ruska, 1934a).*

In classic histology—the study of biological tissues—all kinds of stains are indeed used to enhance contrast in the image. In electron microscopy staining can be done with the large atoms of a number of metals. These atoms will attach themselves to existing structures in the tissue. The strong scattering that they cause, subsequently reveals an image of the structure to which they are attached. This was the first time that the idea of staining was suggested for electron microscopy.

Bodo von Borries was very unhappy with the course that events had taken now, as we have to believe Heinrich Grünewald and Hedwig von Borries. Ruska's comprehensive report on the *Übermikroskop* looks very

[51] Marton believed that he was the first person to realise that scattering was the source of contrast (Marton, 1968). A later exchange between Ruska and Charles Süsskind (Marton's biographer) clarified the situation (see the introduction to the 2nd edition of Marton's book, Marton, 1968); PWH.

much like the intended paper on "innovations of the electron microscope design" in the written agreement between von Borries, Ruska and Freundlich, in which it is stated that the article was to be published by von Borries and Ruska together. Ruska's sister Hedwig writes:

> *"(...) 12 letters were written in which both sides referred to the major joint apparatus study that was agreed upon on January 2, 1933. (...) The sole unfinished task in von Borries' work schedule for December 1933 was to 'compile the apparatus study'. On December 12th, Ruska sent the jointly agreed study (...) to the* Zeitschrift für Physik. *He did not inform von Borries of this until December 21st, by which time the publisher had confirmed acceptance of the article. Deeply upset by this, von Borries considered writing to the publishing house and discussed this with his father over the Christmas period. However, because in recent years there had been several instances of studies lying around for too long and being pre-empted by publications of other scientists, the father and son resolved to let the matter lie. Since the problematic patent applications by Rüdenberg had already been made at the time,[52] and priority disputes with AEG had already commenced, von Borries did not wish his differences with Ruska to become known" (von Borries, 1991; Lin, 1995).*

In 1976, Grünewald had been the first to refer in a historical paper to the written agreement, as well as to its apparent breach by Ruska. At the time, Ruska felt so offended by this remark and other suggestions in the paper, that he sued Grünewald over it. The court case resulted in a settlement, which was published in the same journal in which the original article had appeared. In it, the two parties agree that Ruska had developed the Übermikroskop on his own, and that von Borries and Freundlich have been acknowledged by Ruska in his article of 12 December 1933.[53] However, as far as von Borries is concerned, these acknowledgements are very meagre. In the section about the polepiece lens, a single footnote says:

> *"For the construction design of the coil I have to thank the gentlemen Dr. Ing. B. von Borries and Dr. Ing. M. Freundlich for their valuable advice" (Ruska, 1934a).*

Freundlich received more credit as he was also mentioned in the last sentence of the paper:

[52] By then, von Borries and Ruska knew about the French Rüdenberg patent 737 716, which had been published a year before. Ernst Ruska, Die Entstehung des Elektronenmikroskops (Zusammenhang zwischen Realisierung und erster Patentanmeldung, Dokumente einer Erfindung), *Archiv der Geschichte der Naturwissenschaften* 11 (1984) 525–551, p. 533; Ernst Ruska, The emergence of the electron microscope (Connection between realization and first patent application, Documents of an invention), *Journal of Ultrastructure and Molecular Structure Research* 95 (1986) 3–28, p. 9.

[53] Editorial, Vergleich, *Technikgeschichte* 44 (1977) 360–361.

> *"I thank Herr Professor Matthias for facilitating and supporting this study; I owe thanks to Herr Dr.-Ing. M. Freundlich for his frequent experimental help."*

Of the three men who were involved in the agreement, Martin Freundlich happens to be the only one who has commented personally on this affair in later years. In his historical reflections from 1994, he made clear that he did not agree with Grünewald. Freundlich remarks that shortly after the agreement had been signed, circumstances changed to a very large degree. Because of the political situation, the two joint papers by von Borries and Freundlich were not to appear at all. Meanwhile, von Borries' move to another part of the country had limited his actual contributions considerably, and as a third point he adds:

> *"Besides this, a large and most important part of Ruska's paper shows the experimental results obtained with his new electron microscope; namely the first micrograms ever with a greater resolution than that obtainable with light optics. These results go beyond the scope of the proposed paper and were accomplished by Ruska alone" (Freundlich, 1994).*

Independently of Freundlich, it seems, Lin came to a similar conclusion with regard to the results that had been presented by Ruska, and suggested that the long article on the Übermikroskop design could also be seen as a paper on "very large magnifications with the magnetic electron microscope," which was one of the other intended articles in the agreement, and which was to be written by Ruska alone. I tend to disagree at this point, however, for two reasons. First of all, only the last quarter of the design paper is dedicated to results, while it seems quite logical to include some results that can be obtained with the new instrumental design. Furthermore, Ruska had already published some weeks before his short contribution in *Forschungen und Fortschritte*, which fits the description of an article on "very large magnifications" far better. This would lead to the conclusion that Grünewald's accusation had not been completely unjustified.

Probably, Freundlich provided a better argument, when he referred to his own fate. One could very well argue that the agreement had become void, the moment that it became clear that von Borries was never going to publish with him. If indeed the agreement had been broken, it had already happened well before Ruska submitted his paper on 12 December. And maybe Freundlich was also right about his observation that actually most of the work had been done by Ruska himself. In the 1985 interview, it looks as if Ruska held a certain grudge against von Borries at the time, when he told us:

> *"On the first of April 1933 von Borries went to Rheinisch-Westfälische Elektrizitätswerke as an electrotechnical engineer. So he had taken leave of the business (...).*
> *Money for electron microscopes on a larger scale was not available, and then I had to go to industry as well, but at least I stayed in electron optics."*

After this, there was going to be one more publication by Ruska alone, which was his PhD-thesis on the design of the polepiece lens. Although the thesis had been turned in at the TH on 31 August 1933, it was only submitted to the *Zeitschrift für Physik* on 5 March 1934 (Ruska, 1934b). The thesis does not contain any specific remarks or observations that need to be discussed after everything that has already been said here. The thesis makes it very clear, however, that the polepiece lens was designed exclusively to obtain large magnifications with an electron microscope. It is also worth noting that Ruska acknowledged von Borries and Freundlich together again. In a footnote he thanked both of them for their advice, as well as for their experimental help. In the final sentences, both are thanked for their frequent help and encouragement. From a retrospective point of view, there is certainly a bitter irony to this mention in one breath of Ruska's Nazi friend and his Jewish colleague. Most and for all, it illustrates Ruska's complete lack of interest in—if not contempt for—political issues.

With this last article, the earliest development of the magnetic electron microscope came to its conclusion. In about four years time, Ruska had come a long way: from embryonic electron optics and his first rudimentary electron microscope with Knoll, to the design of the ultimate lens with von Borries, and finally the construction of the Übermikroskop with the occasional help of Freundlich. As far as the scientific community was concerned, it had been an interesting exercise with little practical meaning. The transmission electron microscope could very well have ended here as yet another silly invention. After all, history is crammed with blind-alley contraptions. An example of this is Brüche's cathode-ray aeroplane compass, which was described in Section 2. It seems rather strange, therefore, that at the same time von Borries and Ruska were confronted with very serious competition by Brüche and his colleagues at AEG. An explanation for this apparent contradiction will be provided in the next section.

8. AEG RESEARCH INSTITUTE

It had been on 15 January 1932 that the word *Elektronenmikroskop* appeared in print for the very first time, as we saw in Section 2. So far, it had been the most noteworthy contribution to the development of the electron microscope by Ernst Brüche, the head of the Physics Laboratory of AEG's Research Institute. His short announcement in *Die Naturwissenschaften*, was followed four months later by a much more elaborate paper, titled *Elektronenoptik und Elektronenmikroskop* and published in the same journal on 20 May (Brüche & Johannson, 1932c). It appeared a few months before the third article by Knoll and Ruska, which was going to carry the rather similar title *Das Elektronenmikroskop* (Knoll & Ruska, 1932a). Right from the beginning of their paper, Brüche and co-author Helmut Johannson were going to make it absolutely clear that AEG was not to give up their priority claim easily. The very first sentences read:

> *"It is an old experience that the prerequisites for an experiment, a construction or an invention have been provided by a fundamental study, but nevertheless this valuable material remains unused for years. Suddenly this material is remembered—and the strange thing is that this often happens nearly simultaneously at several places—and the potential of it is recognized. This is also what happened with the electron microscope."*

What they refer to here is an assumed "rediscovery" of the Busch lens theory by both the TH and AEG. This repetition of their claim was followed by a rather loose definition of the concept "electron microscope", simply stating that it is an arrangement that renders strong magnifications by means of electron rays in the way a light microscope does with rays of light. As such it did not differ very much from Knoll and Ruska's definition in their third article, where they said that at least the first stage of the magnification had to be realised by means of electron rays. After that, Brüche and Johannson make some general comparisons between the optics of light and electron microscopes, where they refer to arrangements for two-stage magnifications. The results that they subsequently present, however, are only one-stage magnifications of hot cathode surfaces. The published images are very similar to the two images of a cathode that had been used to illustrate Brüche's first announcement, showing magnifications of up to 150 times made with low acceleration voltages ranging from 200 to 800 V, so less than one kilovolt. The most important experimental feature was the electron lens that they employed, which was an electrostatic lens. It was made of two diaphragms that were sitting very close to the cathode, of which the one that was furthest away, actually acted as anode as well. Cathode, diaphragms and screen were all fitted on riders that could glide over a rail inside a horizontal glass tube, making the whole device a true optical bench.

Especially interesting are the opinions and theoretical insights of the authors. First of all they wish to demarcate their territory by drawing a clear line between their work and that of Knoll and Ruska:

> *"There are two roads that lead to the goal of building an electron microscope, given by the possibility to either construct an electrostatic or an electromagnetic lens. The two roads have been travelled independently by Knoll and Ruska on one side, and the authors on the other."*

The reason to choose the electrostatic design was a practical one. In the paper the authors mention that magnetic electron optics is more complex than electrostatic optics, since a magnetic lens will rotate the electron image, while this is not the case with electrostatic lenses (Brüche & Johannson, 1932c; Lin, 1995). Apart from that, electrostatic lenses are a lot simpler to construct than magnetic ones and therefore more attractive from the engineering and economic point of view. The fact that Knoll and Ruska were already working with magnetic lenses appears not to have been of any consideration. When we asked Ruska in 1985 whether his work could have influenced AEG's choice for electrostatic lenses, he too answered:

> *"No. It was more or less a coincidence. They wanted to perform fundamental electron optical research. And then they noticed that the projection on the viewing screen showed details of which they believed that these came from the cathode, and then they published this image. And then later they have called their microscope an emission microscope."*

Ruska wished to make clear here that AEG was not studying the same objects as he and his colleagues. The publication of 'this image', is a reference by Ruska to Brüche's first short announcement of January 1932. AEG's desire to focus primarily on fundamental electron optical research is repeatedly emphasised by Lin as well. As a consequence of the position that AEG took in this matter, a rather contradictory situation arose in which the research institute of a commercial enterprise dedicated itself to fundamental science, while the actual development of a new instrument took place at the Technische Hochschule, which was an educational institute, funded with public money.

An explanation for AEG's noteworthy role in fundamental research might be found in the character of Carl Ramsauer, the director of the company's Research Institute. Ramsauer was a successful physicist who had achieved fame with his discovery of the Ramsauer effect, also know as Ramsauer–Townsend effect or Townsend effect. Before he made this discovery he had been an assistant for thirteen years to Philipp Lenard, a Nobel

Prize winner. In a memoir, published in 1949, Ramsauer shed more light on the academic climate at AEG, writing:

> *"In AEG, especially in their intellectual leader general director Bücher, I found so much understanding that considerable resources were entrusted to me to conduct, without any unnecessary tutelage, together with my co-workers, important research and development for the benefit of the company and my own pleasure" (Ramsauer, 1949).*

And:

> *"In 1933 I was appointed member of the Board of AEG (. . .). My assignment was of a less material nature. It was my task to enforce the importance of science for industrial engineering, while I received the full understanding and greatest cooperation in particular from the general directors, namely privy councillor Bücher and Professor Petersen—especially, since the intensive fostering of science as well as the foundation of a large research institute had been Bücher's condition when he took over the position of general director, while Petersen regarded research and development his major task."*

This is the same Petersen who published in 1930 the book *Forschung und Technik*, in which Brüche mentioned the words "geometric electron optics" for the very first time.

About his role as director of the research institute, Ramsauer wrote:

> *"In 1931 I was appointed Honorarprofessor [unsalaried professor] to the Technische Hochschule Charlottenburg and I have often put my institute at the disposal of Master and PhD students to carry out their research projects. For the rest, I did not measure my success in terms of the realisation of my own plans—not even my own publications or patents, but in terms of the flourishing of the whole institute. I saw it as my task to supply my co-workers with the encouragement and chance to develop their productive talents in all respects as much as possible, and this I believe, has proved me right, in contrast to Lenard's style of a strict central leadership."*

In earlier pages of his memoir, Ramsauer had paid ample attention to Philipp Lenard's way of supervising his students, which, in his view, left no room at all for personal initiative and was experienced as utterly frustrating by those involved. Having witnessed this, Ramsauer developed a completely different approach, which appears to have been quite similar to the one of Matthias at the High Tension Lab, where Ruska had enjoyed all the latitude he could wish for. As a matter of fact, in 1931 Ramsauer had become an immediate colleague of Matthias, and Rüdenberg too, after his appointment as unsalaried professor to the Berlin polytechnic.

To return to Brüche and Johannson's joint paper after this elaboration on the academic freedom these two authors were allowed to enjoy at their institute, it is also remarkable that they did not say a single word about the high resolution that might be obtained with an electron microscope. It appears they had greater expectations of an electron telescope instead, stating:

> *"Another, very important instrument of geometric optics is the telescope. With an electron telescope we could observe a distant electron-emitting world. Had Piccard flown five times as high, he would have reached the domain of the aurora borealis rays. For now it is a utopian dream, but why should it not be possible to point the electron telescope at this unknown electricity radiation some day? Maybe then we will see the sun spots with their electrical craters and the eruptions of electrical particles!"*

It was already a year since Knoll too had used the telescope as an example of a possible application of electron optics in his Cranz Colloquium. In retrospect, it might be tempting to smile at Brüche and Johannson's rhetoric, but in fact their concluding words reveal something completely different, as I have already suggested in Section 1. The bare thought that you could use an electron microscope to study ultramicroscopical objects was too far-fetched. In contrast to that, this fantasy about studying craters on the sun with electron telescopes, was apparently not that outrageous. After all, one should also keep in mind that this article appeared in a highly esteemed journal, which was once launched as the German counterpart of *Nature*.

AEG's first comprehensive report on the electron microscope is typical of the stand that Brüche would take from here on for the next six years. He was just as eager as Ruska to use the word *Elektronenmikroskop* whenever he could, but the experimental results that AEG published, were invariably one-stage images of the processes that take place at the surfaces of hot cathodes.

Four weeks after the publication of AEG's first detailed report on their electron microscope, Brüche gave a lecture at the meeting of the German Physical Society and the Society of Applied Physics in Berlin on 17 June 1932. This was the meeting at which Knoll had presented the results of the cathode emission study, which he had carried out together with Fritz Houtermans and Werner Schulze from Gustav Hertz' laboratory (Knoll et al., 1932a). Brüche's lecture was published in the September 1932 issue of *Forschungen und Fortschritte*, the same science magazine for a wider audience in which Ruska was going to publish his contribution on the *Übermikroskop* 16 months later. From this magazine we can learn that somewhere between the submission of the first comprehensive report on the electron microscope and the day of his lecture, Brüche must finally have understood the significance of de Broglie's thesis. The very first sentence of the contribution reads:

> *"After de Broglie had assigned a wavelength to the moving electron, a* wave optics *has been developed, which has prompted important experimental research and has accounted for significant theoretical insights. In contrast to this, until now one has not thought of the possibility to develop the* geometric optics *of electrons as well, and consciously apply its principles to scientific progress." [Emphasis by magazine]*

Although the words "de Broglie" are practically the first, this quote does not immediately prove any awareness of de Broglie's significance for electron microscopy. This only happens in the second half of the one page long contribution, when Brüche defined two immediate research goals for electron optics. The first one was the study of the electron emission of tungsten in particular, while:

> *"The second task is of even greater general interest. It involves overcoming the application limit of the light microscope. Since the electron rays that are used for an electron microscope have wavelengths which are several orders of magnitude smaller than the rays that are used for a light microscope, in principal an electron microscope should make it possible to distinguish details that cannot be detected anymore with a light microscope because of their smallness in proportion to the wavelength."*

Brüche, however, was not very convinced about the outcome and adds:

> *"Whether in this case one will succeed in utilising this possibility, and whether no difficulties will arise with regard to intensity and resolution that will limit this ambition beforehand, is something we will have to see."*

It illustrates his awareness that a high resolution can only be obtained with a high intensity of the electron beam, which was very likely to destroy any small object.

Brüche's next publication was submitted to the *Zeitschrift für Physik* on 22 July 1932 (Brüche, 1932b). The title, "The significance of the acceleration field's geometry to the gas concentrated electron beam", already makes it clear that this paper was a return to Brüche's favourite thread beams, or *Fadenstrahlen*. This time he wished to discuss the importance of the shape of the acceleration field for a good 'Fadenstrahl'. He explained that a cathode with a concave surface should be used in order to obtain a proper beam that is not too wide, as the wider it becomes, the more it will tend to blur at its periphery and will even start to diverge. To further intensify the beam, it will help to place a diaphragm as anode at a short distance from the cathode. From here it was a small step to electrostatic lenses, made from a collection of charged diaphragms. It resulted in the first presentation of

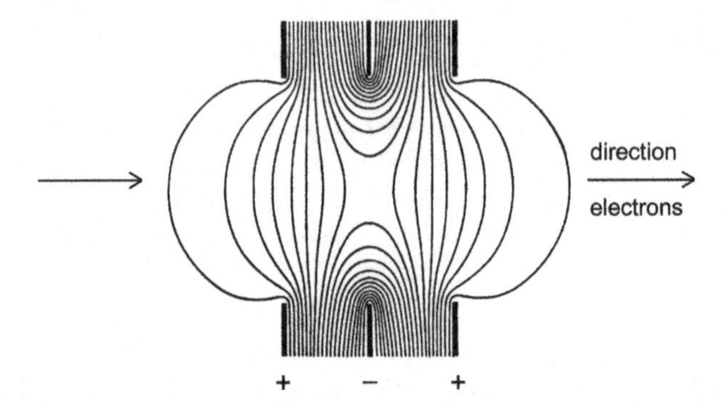

Figure 9 Electrostatic electron lens. The curved lines indicate the electrical field [sic, PWH]. From Johannson and Scherzer (1933, Fig. 2).

AEG's famous *Einzellinse*, which is made of three diaphragms as illustrated in Fig. 9. The outside diaphragms carry the same charge, while the central one carries a charge that is different. Brüche had designed it together with Helmut Johannson, who was going to publish a separate article about it in the next year together with Otto Scherzer (Johansson & Scherzer, 1933). Clearly, the paper has very little relevance to the electron microscope, and Brüche even stressed that he was primarily concerned with electron optics here in order to obtain a size reduction and not an enlargement. This did not deter him from calling the experimental arrangement that he had used for this study an electron microscope since he had used it to produce a ten-fold magnification of the cathode in order to establish certain dimensions of the cathode. Brüche also mentioned 'de Broglie wavelength' again. This time, however, it was a rather superfluous reference to electron speed, while discussing the need for a monochromatic beam. Abbe's equation played no role whatsoever and it certainly had nothing to do with the superior resolution of fast electrons.

Only two days later, Brüche and Johannson submitted another paper, this time to the *Annalen der Physik*. This one carried the fascinating title *Kinematographische Elektronenmikroskopie von Oxydkathoden* and is an extensive report on five time-lapse films that had been presented at the meeting of the German Physical Society and the Society of Applied Physics on 17 June 1932, mentioned above. As we already saw in Section 2, Brüche had made a trip to Norway in 1931, where he had shot a film about the northern lights. Later on, the film camera had come in very handy for recording cathode emissions. He and Johannson had produced films with a length of some 1000 to 2000 frames, of which each frame was exposed for 3 to 6 seconds to the enlarged image of the burning cathode of their electron microscope—the same instrument with one electrostatic lens that they had described in their earlier joint paper that year. It is interesting to note that Brüche and Johannson explicitly referred to Knoll and Houtermans, when discussing some of their results, which illustrates that Knoll and Brüche were actively competing now in the new field of emission microscopy. Finally, the use of the words "electron microscopy" in the title also deserves attention, since it was the first time that it appears in an AEG paper and came shortly after Ruska had done the same (Ruska, 1933a).

Lin tells us that AEG's cathode films were received well by the German physical community (Lin, 1995). According to Gabor, the films were also shown at the 8th German Convention of Physicists in Bad Neuheim, which took place from 20 to 24 September 1932. When Dennis Gabor gave a talk in 1974 to the Eighth International Congress on Electron Microscopy in Canberra, Australia, he talked about AEG's electron optical work:

> *"In 1932 they had not only made many excellent photographs of the activation processes of various hot cathodes, but made a brilliant film of it, which I remember seeing with great admiration at the physicists' convention in 1932 at Bad Neuheim" (Gabor, 1974).*

Brüche and Johannson had indeed been present in Bad Neuheim, where they had given a talk, an abstract of which was published two months later in the *Physikalische Zeitschrift*, called "Some new cathode studies with the electrical electron microscope" (Brüche & Johannson, 1932b). The title shows that the term "electrical electron microscope" had already become familiar in the mean time. Again, the authors present images of cathodes, as the title of their talk already revealed, and as such there is nothing new, at least not from the perspective of this article. There is an intriguing footnote, however, which says:

> *"In this abstract of the presentation, the introductory examples have been left out, among which the application of magnetic and electrical deflection to electron optical images, and their combined application. These examples (...) will be discussed in more detail at a later stage. We should like to add, however, that based on our experiments, the combination of an electrical objective system with a magnetic ocular appears to us to be the way to realise high electron microscopical magnifications."*

By "ocular" they meant the second lens in a two-stage magnification, which can be likened to the eye piece of a light microscope, also known as ocular. This footnote is remarkable for the fact that the authors suddenly open the door to magnetic lenses here the very moment that high magnifications are concerned. From the AEG Research Institute's yearbook

1931/32 (Brüche, 1931/32)—which was published only after May 1933—it can be learned that as a matter of fact the experiments with the magnetic lens had failed, since the resolution of the electrical objective had not been good enough. From this same contribution it can also be learned that the design of the magnetic lens had been copied from Knoll and Ruska's long article that had appeared in early 1932 in *Annalen der Physik*.[54] In this paper a drawing of an encased coil without polepieces had been published for the first time. It is rather telling of course that Brüche and Johannson had to resort to Knoll and Ruska's encased magnetic coil the moment that they wanted to achieve higher magnifications.

When looking at the articles that were published by the AEG Research Institute, it is clear that Brüche preferred to stay on the safe side and to invest in electron optical research in general with an initial emphasis on the rapid success that was gained with emission studies. Further evidence for this comes from two more papers that were published by AEG in 1932. On 20 November 1932, Helmut Johannson and his colleague Otto Scherzer submitted a paper to the *Zeitschrift für Physik* with the title *Über die elektrische Elektronensammellinse* (About the electrical electron lens) (Johannson & Scherzer, 1933). It dealt exclusively with the *Einzellinse* that Brüche had already presented a few months before. The paper is a theoretical study essentially, which is even regarded as fundamental, since the authors showed by theoretical means that an electrical electron lens, such as an Einzellinse, can never be concave, regardless whether the middle electrode of the lens is negative or positive.[55] In light optics, concave lenses are widely used to correct optical errors, and therefore the impossibility of constructing concave electron lenses was a serious setback to Brüche and his team, since electron lenses tend to show serious optical errors that limit the performance of the electron microscope. Johannson's and Scherzer's important finding followed from a new formula for the focal distance of an electrical

[54] Ernst Brüche, Geometrische Elektronenoptik, in: *Jahrbuch des Forschungs-Instituts der Allgemeinen Elektricitäts-Gesellschaft* 3 (1931/32) 111–124, refers on p. 115 to Max Knoll and Ernst Ruska, Beitrag zur geometrischen Elektronenoptik (I and II), *Annalen der Physik* 404 (1932) 607–661.

[55] A concave lens for light rays has a negative focal distance, meaning that it will diverge light rays instead of converging them. Proof that this handicap applied to all electron lenses was finally provided in 1936: Otto Scherzer, Über einige Fehler von Elektronenlinsen, *Zeitschrift für Physik* 101 (1936) 593–603. [This is slightly misleading, it should be read in conjunction with works on electron optics; PWH.] This does not apply to the lenses made of wire mesh, like the ball condenser that Ruska constructed in 1930 for his second assignment.

lens, which had actually already been presented at the physicists' convention in Bad Neuheim. Johannson had mentioned it during the discussion after their presentation, and the formula had therefore already been published in the *Physikalische Zeitschrift*. There, however, the impossibility of concave electron lenses was not yet pointed out in so many words.

To establish whether their new formula was correct, Johannson and Scherzer tested the Einzellinse in an "electron optical microscope system" as they called it. Again it was a true "optical bench", to quote the authors themselves, with the cathode, object, the single lens and the viewing screen fitted on riders on a horizontal rail inside a glass vacuum tube. The object was a bronze wire mesh. As it had been their conclusion that it does not matter whether the middle electrode of the lens will be positive or negative, they also tried to apply alternating current to the system instead of direct current. The experiment showed that their theory was completely correct. Finally, it is worth mentioning again that in this paper too Max Knoll is explicitly acknowledged for his work on electrostatic lenses.[56]

The article by Johannson and Scherzer was immediately followed on the next page of the journal by a paper by Scherzer alone, called *Zur Theorie der elektronenoptische Linsenfehler* (Scherzer, 1933). Probably the most noteworthy remark is made by Scherzer in the introduction, where he observes that electron optics will have to be worked out just as thoroughly as light optics, if there is any advantage to be gained from the very short waves of electron rays with regard to microscopy. One of his major conclusions was that the chromatic aberration of electrical lenses will increase with lower voltages. This must have been another setback for Brüche who did not really appreciate high voltages. It needs no explanation that this paper too was of a very theoretical nature.

After these papers it became comparatively quiet at the AEG Research Institute as far as electron microscopy was concerned. As already mentioned, a contribution by Brüche on geometric electron optics appeared in the AEG Yearbook for 1931/32 at some time shortly after May 1933. It is here that Brüche introduced the word *Übermikroskop* for the very first time:

> *"In the further development of imaging systems, the work is directed towards the distant goal of the "Übermikroskop", that is, a microscope which is still able to resolve—thanks to the smallness of the electron wavelengths—where a light microscope has reached its fundamental limit by far."*

[56] Helmut Johannson and Otto Scherzer, Über die elektrische Elektronensammellinse, *Zeitschrift für Physik* 80 (1933) 183–192, p. 183, footnote 2.

Also this reference to an electron microscope in the modern meaning of the word, is immediately followed by critical remarks. To start with, Brüche has his doubts about, what could be called, a high resolution emission microscope with which one might study microscopic details of self-emitting objects. First of all, he foresaw a problem here with the optical errors of the electron lens, which could not easily be solved because of the theoretical impossibility to construct a concave lens. In the second place it would be difficult to obtain a sharp image of an object, since the surface of the object will be very rough relative to the size of the details that one would like to see. As a consequence it would be impossible to focus properly. Apart from that, Brüche noted that the irregularly shaped electrostatic field of an emitting surface will distort the field of the objective lens. Finally additional chromatic errors could be expected if one would like to image "non-emitting objects" which was a reference to transmission electron microscopy, as we call it today. Brüche expected that the speed of the electrons in the beam would become differentiated after their passage through the object.

So, in total we have four critical notes here. Brüche could have applied the first general note about optical errors to transmission images just as well. It was probably the most legitimate of the four, and as a matter of fact it is a major reason why the classic [uncorrected; PWH] transmission electron microscope cannot go beyond the 1 Å limit. With his second note about the focusing problems, Brüche proved to be unaware at the time of the excellent depth of field of a proper electron lens, thanks to its very small aperture. The third note about the lens distortion certainly made sense, but high resolution microscopy of emitting surfaces was never to become an issue of major concern. Brüche's fourth note on the differentiation of electron speed, finally, was based on the assumption that images would be the result of inelastic scattering of electrons by the object. Only then you will have a loss of energy and a subsequent loss of speed of a portion of the electrons.

A few months later, an abridged version of Helmut Johannson's PhD thesis was submitted to the *Annalen der Physik* on 25 August 1933, called *Über das Immersionsobjektiv der geometrischen Elektronenoptik* (Johannson, 1933). It dealt with the type of electrical lens that had already been presented a year before in the first joint paper by Brüche and Johannson. The experimental results that Johannson presented to demonstrate the performance of his lens were all images of cathodes again. Despite its rather technical nature, this paper too contains a few casual references to future possibilities of electron microscopy. When discussing the perceived problem of focusing on the irregular surface of a cathode with substantial protrusions, Johannson writes:

> *"It is conceivable, however, that one of the final goals of electron optics—namely, the construction of an* Übermikroskop *with which one theoretically may even resolve objects the size of a molecule (...)—will fail because of the roughness [irregularity] of the objects that are to be imaged."*

As we just saw, this argument did not take the depth of field of a proper electron lens into account. More importantly, however, it is another proof of AEG's ambivalence. Brüche had already been critical about the necessary intensity of the electron beam and the consequences of this for an object in the *Forschungen und Fortschritte* edition of September 1932. In the AEG Yearbook for 1931/32, Brüche had come up with four more problems, and now it was Johannson who wondered whether it would be possible to focus properly. Even so, Johannson pointed out on the next page that one of his images was so sharp that it would not be a big step to surpass the resolution limit of a normal light microscope.

Altogether, it remains a remarkable situation that AEG was investing this much energy in rather fundamental electron optical research and in emission microscopy as its main application. According to Lin, the company believed that emission microscopy was going to pay off soon, because of the importance of cathode-ray tubes for the fast-rising oscilloscope and television industry. Meanwhile, *Übermikroskopie* was regarded a distant goal which would be difficult to achieve, considering the many technical difficulties that were foreseen on theoretical grounds, as can be read in Lin's study as well (Lin, 1995). As Brüche was the head of the Physics Laboratory of the Research Institute, he appears to have been primarily responsible for this ambivalent attitude. There is no evidence that Ramsauer had any significant influence on this. In his 1949 memoir of 32 pages, Ramsauer mentions the development of electron microscopy at the AEG Research Institute at two places, but only very briefly, without any comment whatsoever, and in both cases in immediate connection to Brüche's assignments (Ramsauer, 1949). In contrast to this, he spends three full pages on his contributions to the development of targeting techniques for German antiaircraft artillery during World War I, and many more pages on his years with Lenard, not hesitating to make numerous critical notes.

In the end, Brüche's real commercial breakthrough was an unexpected one. On 16 September 1933, Brüche submitted a paper to the *Zeitschrift für Physik*, titled *Elektronenmikroskopische Abbildung mit lichtelektrischen Elek-*

tronen (Brüche, 1933). This article was to become the classic first study of photoemission electron microscopy (PEEM). The principle as such had already been patented by Knoll, Houtermans and Schulze, as we saw in the previous section, but they had not made any mention of it in their own papers. Only in the third Knoll and Ruska paper is a short reference made to the theoretical possibility to use photoelectric electrons as a radiation source for imaging (Knoll & Ruska, 1932a). The arrangement that Brüche describes is very similar to the one that Knoll and Houtermans had constructed, as it was also fitted with an electromagnetic lens. Just like the ones that were described by Knoll and Houtermans, Brüche's magnetic lens was encased in iron, with a narrow slit in the iron at the inside of the coil, and fitted at the outside of the tube, meaning that the tube wall ran through the passage in the lens. The only difference is that Brüche used one instead of two lenses. Since Brüche has acknowledged in the AEG 1931/32 yearbook that he had copied Knoll and Ruska's design of an encased magnetic coil, it seems fair to conclude that the same must apply to this lens for his photoelectric microscope. It is also noteworthy that he was applying far higher acceleration voltages now, up to 35 kV, which might very well explain why he needed to use a magnetic lens, since electric lenses were not capable of operating with such high voltages then.

In his experimental setup, Brüche had used a mercury lamp and a quartz lens to project a very strong beam of light on a zinc plate. The photons of the light would energise electrons in the zinc plate, which were subsequently focused on the viewing screen by the lens. With this apparatus Brüche had been able to make a quite impressive magnification of a pattern of holes and the letters Zn, which he had engraved in the plate.

With this photoelectric application Brüche had finally found an electron optical application from which AEG would benefit greatly. Photoelectricity works at all sorts of wavelengths. If a photoelectric material is used that is sufficiently sensitive to infrared for example, night vision gear can be built. Infrared is emitted by warm objects, like humans, working engines and heated buildings. The photoelectrical material will translate the infrared image into an electron image and with a fluorescent screen the electron image is translated into a visible light image. This way anything can be seen at night that radiates heat sufficiently. It will need little explanation that the German military was extremely interested in this ability to see at night, certainly after Hitler had come to power in the spring of 1933. Nor is it surprising that AEG published very little about photoelectric imaging in subsequent years. In a personal record from 1949, Brüche wrote about this period that,

> *"The army had declared night vision secret. Today it may be revealed that in the following years considerable results were obtained with the technology to see at night" (Lin, 1995).*

This brings a kind of natural end to AEG's early contributions to the history of the electron microscope. These contributions as such have shown to be rather contradictory in this period. On one hand there is a continuous and frequent use of the words "electron microscope" and "electron microscopy." On the other hand, all of AEG's experimental results concern one-stage magnifications of emitting cathodes. Subsequently, Brüche is the one who introduced the word Übermikroskop, and he and his co-workers start to refer to the distant goal of overcoming the resolution limit of common microscopes. At the same time, however, they refrain from making efforts in this direction and limit themselves to all sorts of arguments why it was going to be very difficult to achieve this goal. And so we end at this point with the situation that Ruska had wanted to continue to develop transmission microscopes, but could not, while AEG could, but chose not to.

9. BELGIUM

On 4 June 1934, Ernst Ruska received a letter from Brussels, which contained a surprising message:

> *"Dear colleague, Since approximately one and a half years I have been occupied with putting an electron microscope into operation and applying it to biological subjects. I have always followed your articles with great interest and wanted to contact you for some time,—but first I wanted to wait for my first articles on this topic to appear. As I have to travel to Hungary suddenly, I don't want to delay any further, but wish to ask you the following: I would have the opportunity to travel through Berlin on my way back. I would be very pleased if I could meet you at that occasion and further discuss with you some questions that interest me."*[57]

The letter was signed by Dr. L. Marton from the Department of Applied Sciences of the Free University of Brussels, Belgium—"free" basically meaning "state-independent" here. From this moment on in 1934, Marton would become a prominent character in the history of electron microscopy,

[57] Ladislaus Marton, letter dated 4 June 1934, personal archive Ernst Ruska. A copy of it is in my possession.

as he became the most industrious advocate of the magnetic transmission electron microscope outside Germany before World War II. Because of the international role that he was going to play, Marton may even be designated ambassador of early electron microscopy. Actually, this is not how Marton himself saw it. He preferred to regard himself in the first place as discoverer and inventor, a view that would give rise to incidental misunderstandings and disagreements.

Ladislaus Lászlo Marton was born in Budapest, the capital of Hungary, on 15 August 1901.[58] This made him some five years older than Ruska. Being of Swiss–Hungarian descent, he attended high school in Hungary and studied physical chemistry at the University of Zürich in Switzerland, where he received a doctorate in infrared spectroscopy. Since Zürich is in the German speaking part of Switzerland, it is clear that Marton's command of German must have been more than sufficient to understand the German papers about electron microscopes. After his return to Hungary, he worked for several years at Tungsram Lamp Company, a large manufacturer of light bulbs. At some stage he realized that he had more profound aspirations, and in 1930 he moved to Brussels where he became a research associate with Professor Emile Henriot at the Free University. Initially he was going to work on photoelectricity, X-rays and infrared (Süsskind, 1985). Going by his letter to Ruska, it must have been late 1932 that he turned to electron microscopy. In his "Early History of the Electron Microscope", which was published in 1968, Marton wrote:

> *"I was fortunate enough to have a department chairman who coupled a broad outlook with a fine sense of experimental possibilities. Professor Émile Henriot (1885–1961) became a good friend and I could freely talk over my ideas and problems with him. He immediately encouraged me and, although the laboratory resources were rather meagre, gave me the means for a first try. The result was my very first primitive electron microscope, which was in operation by the beginning of December 1932."*

This first microscope was a horizontal brass tube, with a coil of loosely wound wire around it, which acted as a lens. With the magnetic field of this coil he managed to obtain images of the cathode, in this way establishing that he was able to get results. These images were never published at the time, since Marton did not believe that he would be able to compete with a well-equipped industrial laboratory like Brüche's Research Institute at AEG. Nevertheless Marton had got a taste for electron microscopy, and

[58] After his emigration to the US in 1938, he became known as Bill Marton.

as we can learn from the quote above Henriot gave him room to start exploring the new field. In this respect, Henriot's role seems to be comparable to that of Matthias in the case of von Borries and Ruska, and Ramsauer in the case of Brüche.

The explicit reference to the application of the electron microscope to biological subjects, which Marton makes in his letter to Ruska, is interesting since he published his first paper on this subject in 1934. Marton's earliest paper on the electron microscope itself, however, appeared halfway through 1933 in a Dutch journal, called *Wis- en Natuurkundig Tijdschrift*. It was a review of electron optics, titled *Meetkundige optica der electronen* (Geometrical optics of electrons) (Marton & Nuyens, 1933). He had written it together with mathematician Maurice Nuyens, also from the Free University of Brussels. The paper was a quite elaborate discussion of recent achievements, which started with three pages in which the electron optical work of Hans Busch was explained. The remaining nine pages were used to give an overview of the German work on electron optics in general. By the end, attention was also paid to possible applications like electron spectroscopes, microscopes and telescopes. It is striking, however, that the authors only discuss the transmission microscope and its potentially very high resolution, and do not mention the far more popular emission microscopy at all. The paper is illustrated with Knoll and Ruska's two photos of a molybdenum wire mesh, which they made to compare the performance of the electron microscope with the light microscope. There are no images at all of cathodes. Without doubt, the significance of this article is therefore that it was the first time that international attention was paid to magnetic transmission electron microscopy explicitly.

Despite this, it must be admitted that the paper could have been more accurate. Among the minor details are a number of typing errors and misspelled names—three times the Braun tube is called Brown tube for example. More serious, however, are the careless references. The article contains 11 illustrations, apart from a table. For nine of them a source is given: one is copied from Busch, one from Brüche, two from Ruska and von Borries, two from Ruska and Knoll, and three from other authors. The two remaining figures, numbered 2 and 8, have no source and so give the misleading impression that they were drawn by the authors themselves. However, if you combine these two illustrations, the result is Fig. 25 from the large Knoll and Ruska article *Beitrag zur geometrischen Elektronenoptik*. One can even tell that it was copied by hand, because of minute differences. This *Beitrag* article of Knoll and Ruska, however, is not mentioned in the article—only their third article appears in the list.

A similar thing occurs with the estimation of the feasible resolving power of the electron microscope. The authors provide examples, which are exact copies of the numbers that Knoll and Ruska used in *Das Elektronenmikroskop*, without saying so. In their article they take the same wavelength of light as example of the resolution in traditional microscopy. Subsequently this is compared to the same two different electron wavelengths that Knoll and Ruska took, resulting in the same resolutions: 15 Å with 1500 V electrons and 2.2 Å with 75,000 V electrons, while applying the same numerical aperture.

When discussing experimental results, Marton and Nuyens are not very specific about what had been achieved by whom. Although the paper only focused on transmission microscopy, Knoll, Ruska, Brüche and Johannson are equally acknowledged together, which gives the idea that all four men were doing the same thing. Some might argue that this will indeed have been the general impression to many outsiders at the time—if such was indeed the case, this Dutch paper may very well serve as evidence. However, it is also clear that the mentioned inaccuracies will not have been intentional. To the extent that Marton was responsible for the article, it is more likely that they are illustrative of Marton's colourful character. From Nicholas Rasmussen's study (Rasmussen, 1997), Marton certainly comes forward as a complex and not very practical person, to phrase it mildly.

Despite this slightly uneasy start, Marton should certainly be recognised for his pioneering role in electron microscopy. Especially his very early biological application of the instrument is of historical significance.

10. UNITED STATES

1933 was also the year in which the first elaborate publication on electron optics appeared in the United States. It was written by Vladimir Kosma Zworykin from the RCA Laboratories in Camden, New Jersey, the same man who in 1940 would launch the first commercial transmission electron microscope in the US. In 1933, however, the transmission microscope did not have his interest at all.

In popular American history, Zworykin is certainly a legendary figure, often depicted as father of American television, because of his very early involvement with cathode-ray tubes, television and the television industry.

He was born and raised in Russia, where he studied electrical engineering with Professor Boris Rosing at the Institute of Technology of Saint Petersburg. As early as 1907 Rosing had invented "a device for the electrical remote transmission of images by means of cathode rays" as the first sentence of his German patent says.[59] His design was based on the idea of the German Paul Nipkow to scan an image line by line, and then transmit these lines to a device that will assemble the lines into an image again.[60] In 1912, at the age of 23, Zworykin continued his study at the College de France in Paris for two years. In 1919 he settled in the USA, where he invented the so-called iconoscope, which became one of the two popular types of television cameras, together with the system that was invented by the American Philo Taylor Farnsworth. From 1929 onwards Zworykin was the driving force behind the development of television at the Radio Corporation of America (RCA), and from approximately 1938 onwards he took on the same role in the development of American electron microscopy (Rajchman, 2006).

On 24 February 1933, Zworykin presented a talk on electron optics to the Optical Society of America, which was published in the May 1933 issue of the *Journal of the Franklin Institute* (Zworykin, 1933). The paper carried the straightforward title "On Electron Optics", which is nearly identical with the title of the talk that G.P. Thomson had delivered in December 1931 (Thomson, 1931). Its content was completely different, however, clearly showing that it was written by an engineer, interested in applications and not in abstract insights. As a matter of fact, it recalls in several ways the paper by Marton and Nuyens that was published practically simultaneously. Like Marton's, Zworykin's paper is also a general introduction to applied geometrical optics, and so it also begins with a thorough explanation of the lens theory of Hans Busch, which is very well written in Zworykin's case. However, as soon as he focuses on applications, the differences become very apparent, and it is not difficult to recognise that a man whose main interest at that time was television was speaking here.

Evidently, his primary concern was a sharp dot on the screen, so his interest in electron optics was completely focused on the possibilities that it offers to reduce the size of this dot. It is logical therefore that he shared

[59] Boris Rosing, Verfahren zur elektrischen Fernübertragung von Bildern, *German Patent 209320*, priority date of 26 November 1907 and published on 24 April 1909.

[60] Paul Nipkow, Elektrisches Teleskop, *German Patent 30105*, priority date of 6 January 1884 and published on 15 January 1885.

Brüche's opinion that the rotation of the electron image in an electromagnetic field will complicate the situation. This problem is especially worsened, he adds, when deflection plates are added to steer the electron beam. Because of the rotation, it becomes difficult to predict where the beam will go upon deflection. Therefore the abstract that precedes the article already concluded with:

> *"In focusing electron beams, both electrostatic and electromagnetic methods have been used extensively. The electrostatic method, however, seems to be preferable, especially when the beam is to be deflected. Precautions should be taken not to destroy the focusing of the beam during deflection."*

After having dedicated about a third of his paper to Hans Busch, he turns to the second Knoll and Ruska paper, which appeared in *Annalen der Physik*. It was the paper in which they presented the first two-step magnifications and used the words "electron microscope" for the first time, but Zworykin paid no attention at all to the magnifications (Knoll & Ruska, 1932b). He was only interested in the electrostatic ball condenser that Ruska built for his final assignment as a master's student, and which was described in the paper as well. Zworykin's exclusive focus on television-related issues becomes especially clear in his final section on applications, which is titled "Electron microscope" and which might very well be the first time that these words appeared in print in North America. The first sentence of the section reads:

> *"A very interesting and direct application of electron optics is the case of the so-called 'electron microscope' described by E. Bruche and H. Johanson [sic]."*

A footnote contains a rather unclear literature reference to an article in *Annalen der Physik* in November 1932, which must be Brüche's and Johannson's paper on the time-lapse films that they had made of the cathode emission (Brüche & Johannson, 1932a). The note also refers to an earlier paper by Brüche on thread beams (Brüche, 1932c), which has no immediate relevance to the electron microscope or emission microscope. In the same note, he also refers to the second article by Knoll, Houtermans, and Schulze (1932b). Since Knoll and Houtermans also studied cathode emissions, although with magnetic lenses, it already illustrates that in this case it was the cathode studies that had Zworykin's main attention.

This is even more evident from some experimental results that Zworykin had obtained himself. In the paper he reproduced a few images that he made, and it can be learned that he had even done experiments with secondary emission. For this purpose he had covered a nickel surface with carbon, in which he had scratched the letter A. Upon irradiating the surface with electrons, only the A-shape would emit secondary electrons, which he was able to project on the viewing screen. It is quite surprising that Zworykin must have done this experiment at the same time that Ernst Ruska made his own with secondary emission (Ruska, 1933a). Considering Ruska's submission date, one should even conclude that Zworykin was earlier, although he did not present his reflections as "microscopical imaging", as Ruska did. Actually, Zworykin did not even see the electron microscope as a true instrument, stating:

> *"It is not even necessary to construct a special device for observation of the cathode surface, since practically every high vacuum cathode oscilloscope will serve the purpose. It is usually necessary only to readjust the accelerating potentials and to lower the cathode temperature (to reduce space charge) in order to receive a large image of the cathode."*

From a retrospective point of view it is certainly noteworthy that Zworykin did not pay any attention to the possibility of surpassing the resolution limit of the light microscope with electron optics, even when he did adopt the term electron microscope. However, this might very well have to do with an apparent lack of knowledge of quantum physics, as we have already seen before in Berlin. In his discussion of Hans Busch's work, he remarked:

> *"Hamiltonian classical mechanics shows an analogy between the behavior of small particles in motion and the action of light rays. A particular case of this general analogy is the motion of charged particles in a homogeneous electric field."*

There is no mention at all of de Broglie or electron wavelengths in his paper, and when discussing his own attempt to image a wire mesh, he informed his audience that he had obtained "a sharp *shadow* of this mesh on the fluorescent screen."

Why Zworykin had not yet noticed the de Broglie wavelength in the literature is unclear. The news should have reached him by the time that he gave his talk. Despite this, Zworykin's paper reveals that engineers in the US were well informed about German developments. As a matter of fact, this was certainly the case the other way round as well. The paper on photoelectric imaging which Brüche submitted in September 1933, already contained a reference to Zworykin's paper (Brüche, 1933).

11. FRANCE

Finally, attention should be paid to the work of Jean-Jacques Trillat in France, mainly for the sake of completeness. The reason is the existence of a short memoir, written by Trillat and published in 1981 in the book *Fifty Years of Electron Diffraction*, which was an initiative of the International Union of Crystallography (Trillat, 1981). Trillat had obtained a PhD in Paris at the laboratory of Louis de Broglie's brother Maurice, who was an X-ray physicist. Trillat worked there from 1924 till either 1932 or 1933, and as a matter of fact he was even a co-student of Louis de Broglie, since the latter frequented his brother's laboratory while working on his famous thesis.

In the memoir from 1981 Trillat wrote:

> *"In 1932 I left the de Broglie laboratory to take up the Chair of Physics at the Faculty of Science at Besançon. (...) This laboratory was very well equipped for X-ray and electron diffraction and in 1933 I was able, with the aid of my assistant M[onsieur] Fritz, to construct the first French electron microscope, an apparatus which, though rudimentary, aroused a keen interest among the physicists."*

A photo of this apparatus from 1933 was included in a review of the book, which was published in *Science* in 1982 (Howie, 1982).

Unfortunately, Trillat did not supply any additional details and did not publish anything about this first French microscope. A bigger problem, however, is the existence of another memoir, also written by Trillat, but published 19 years earlier in the book *Fifty years of X-ray Diffraction*, which was also an initiative of the International Union of Crystallography (Trillat, 1962). In this earlier memoir, Trillat wrote that he moved to Besançon in 1933 and built the first French electron microscope in 1935.[61] From this earlier memoir it also becomes clear that the full name of Trillat's assistant is Réné Fritz. This is relevant, since Réné Fritz wrote an introduction on the electron microscope in his quality of at the Faculty of Sciences of Besançon, which was published in the *Revue générale des Sciences pures et appliqués* in 1936 (Fritz, 1936). The article's short bibliography does not reveal any link to electron microscopical work being done in Besançon, nor does Fritz make any mention of it in the text. For this reason, for the fact that Fritz only wrote the article in 1936 and for the fact that Trillat's older memoir is more detailed, it makes more sense to go by the later date of 1935 for the first French microscope.

[61] Jean-Jacques Trillat, Some Personal Reminiscences, in: P.P. Ewald (Ed.), *Fifty Years of X-Ray Diffraction*. Utrecht: Oosthoek's Uitgeversmaatschappij, 1962, pp. 662–666, p. 663 and p. 665 respectively.

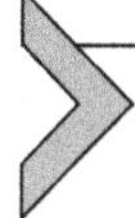

12. MATERIALISING THE IDEA: SUMMARY AND CONCLUSION

Again it will be helpful to start with a chronological summary of major events in order to be able to draw some conclusions from this part and use them to further develop a general overview of the early history of the electron microscope. The summary of this part is presented as the following list of 33 events.

14 February 1932 Fritz Houtermans from the Physics Institute at the Berlin Polytechnic mentions de Broglie for the first time in relation to electron microscopy during a lecture in which he presented the results of experiments, conducted together with Knoll and Schulze (Knoll et al., 1932a).

17 March 1932 Von Borries and Ruska from the High Tension Lab at the Berlin Polytechnic have their patents filed on the polepiece lens and the intermediate screen.[62]

17 March 1932 Knoll, Houtermans and Schulze have their patent filed on the photoelectric electron microscope.[63]

1 April 1932 Knoll leaves the Technische Hochschule for a better paid job in the television industry (Telefunken). Von Borries takes over his position.

13 April 1932 Ruska presents his first research plan to professor Matthias for the development of a magnetic transmission electron microscope, which will include polepiece lenses and the intermediate screen. It is the first time that the words "electron microscopy" appear in writing (Ruska, 1979).

22 April 1932 Von Borries and Ruska (1932a) submit their first joint paper. The subject is the space charge lens.

20 May 1932 First article by Brüche and Johannson (1932c) from the AEG Research Institute on electron optics and the electron microscope with a

[62] Bodo von Borries and Ernst Ruska, Magnetische Sammellinse kurzer Feldlänge, German Patent 680284, filed on 17 March 1932 and granted on 3 August 1939; Bodo von Borries and Ernst Ruska, Anordnung zur Beobachtung und Kontrolle der im Strahlengang eines Elektronenmikroskops mit zwei oder mehr elektronenoptischen Vergrösserungsstufen auftretenden elektronenoptischen Bilder, *German Patent 679857*, filed on 17 March 1932 and granted on 15 August 1939.

[63] Max Knoll, Fritz Houtermans and Werner Schulze, Elektronenmikroskop, bei dem Elektronen aussendende Substanzen in vergrößertem Maßstabe abgebildet werden, *German patent 679330*, priority date of 17 March 1932, granted on 13 July 1939.

first presentation of an electrostatic lens made of diaphragms. The paper contains one-stage images of cathode emissions that were obtained with it at the AEG Research Institute.

16 June 1932 Knoll and Ruska (1932a) submit their third joint paper, in which they apply de Broglie's equation for the first time in order to estimate the electron microscope's resolution, and in which they also present the polepiece lens for the first time.

17 June 1932 Knoll shows the results of the study by him, Houtermans and Schulze on cathode emissions with a two-stage magnetic microscope at a meeting of the German Physical Society and the Society of Applied Physics (Knoll et al., 1932b).

17 June 1932 At the same meeting Brüche and Johannson (1932a) present their time-lapse films of cathode emissions.

17 June 1932 At the same meeting Brüche (1932d) makes his first explicit reference to de Broglie's equation.

22 July 1932 Brüche (1932b) submits a paper in which he presents the *Einzellinse* (single electrostatic lens).

24 July 1932 Brüche and Johannson (1932a) submit a comprehensive report on their time-lapse films of cathode emissions.

9 August 1932 Knoll et al. (1932b) submit their comprehensive report on the cathode emissions of a two-stage magnetic microscope.

1 September 1932 In a paper for a non-specialist audience, Brüche (1932d) pays attention to the theoretical possibility to surpass the resolution limit of light microscopy, but adds that it remains to be seen whether this is feasible in practice.

20–24 September 1932 Brüche and Johannson (1932b) mention during the Eighth German Convention of Physicists in Bad Neuheim that they had used an encased magnetic lens in order to obtain high magnifications.

20 November 1932 Johannson and Scherzer (1933) submit the paper in which they show that there is no way to construct proper concave electrical lenses.

2 January 1933 Von Borries, Ruska and Freundlich sign their mutual agreement.

24 February 1933 Zworykin from RCA gives a talk on electron optics, which results in the first American paper that pays attention to emission microscopy (Zworykin, 1933).

1 March 1933 Von Borries leaves the Technische Hochschule for a better paid job outside Berlin and Freundlich takes over his position.

13 March 1933 Ruska presents his second research plan to Matthias, this time with the specific goal to construct a magnetic transmission electron microscope that will enlarge at least 10,000 times (Lin, 1995).

7 April 1933 Von Borries and Ruska (1933b) submit their second joint paper, which is dedicated to transmission images of aluminium foils. It is the first time ever that published results were obtained with a polepiece lens.

2 May 1933 Ruska submits a paper on the imaging of reflected electrons in which he publicly refers to his intention to realise 10,000-fold magnifications.

12 May 1933 Ruska (1933b) submits a paper on the theory of focussing electron beams that are generated with large cathodes.

May or June 1933 Brüche (1931/32) casts the word *Übermikroskop* (supermicroscope) and gives four arguments why it might be difficult to achieve results with it.

Summer of 1933 Marton and Nuyens (1933) from the Free University in Brussels publish in a Dutch journal the first paper that pays international attention to magnetic transmission electron microscopy and the high resolutions that might be achieved.[64]

25 August 1933 Johannson (1933) submits an abridged version of his PhD as a paper in which he describes an electrostatic lens.

31 August 1933 Ruska completes his PhD thesis on the polepiece lens, which is submitted for publication on 5 March 1934 (Ruska, 1934b).

16 September 1933 Brüche (1933) submits a paper on photoelectric electron microscopy.

23 November 1933 Ruska (1934c) submits a paper to a journal for a nonspecialist audience in which he presents his Übermikroskop for the first time together with magnifications of up to 12,000 times. It is the first time that biology and medicine are mentioned as potential application fields.

[64] Ladislaus Marton and Maurice Nuyens, Meetkundige optiek der electronen, *Wis- en Natuurkundig Tijdschrift* 6 (1933) 159–170.

1 December 1933 Von Borries and Ruska have their third joint patent filed, which describes a method to intensify the electron beam.[65]

1 December 1933 Ruska leaves the Technische Hochschule for a paid job in the television industry.

12 December 1933 Ruska submits a final paper on the construction of an improved magnetic transmission electron microscope with polepiece lenses, intermediate screen, object chamber, water cooling and steel column. It contains a first reference to elastic scattering as an explanation for image formation, and the first suggestion to use certain metals as a dye for staining objects (Ruska, 1934a).

Evidently this list is slightly arbitrary, but nevertheless I believe it is a proper reflection of the historical landscape at the time, based on my assumption that no events of major significance have been omitted. If, indeed, the list can be regarded as sufficiently accurate, the first observation will be that events were dominated by two persons in particular, Ernst Ruska and Ernst Brüche. Ruska's name appears in 14 out of 33 listed items, while 9 items refer to Brüche. Their respective co-workers Max Knoll, Bodo von Borries and Helmut Johannson share a third place with 6 items each. The absolute dominance by a representative of the High Tension Laboratory is reflected by the laboratory itself. The High Tension Lab as such is represented with 18 items, if we include Knoll's activities, when he was still employed by the lab. Eleven items are linked to the AEG Research Institute. The other four are associated with the laboratory of Gustav Hertz, Marton's laboratory in Brussels and the RCA lab in the USA.

The dominance of Ruska and the High Tension Lab seems to be in good agreement with the general impression that we already received from the previous part. In 1931, the two main players were the High Tension Lab at the one hand and Siemens at the other of which only the High Tension Lab was doing actual research. Although Brüche was already very present then, his contributions were predominantly of a linguistic character. In 1932 and 1933, Siemens and its representative Rüdenberg rapidly fade from the stage, but their role is only partially taken over by AEG. Meanwhile, linguistics remains Brüche's territory and he even manages to add the invention of the word Übermikroskop to his trophies. At the same

[65] Bodo von Borries and Ernst Ruska, Anordnung zur Erzeugung eines parallelen oder kegelförmigen Kathodenstrahlen großer Dichte, *German Patent 692335*, filed on 1 December 1933 and granted on 23 May 1940.

time, his real interest is an application of electron optics that could be called emission microscopy, and which is mainly concerned with the magnified imaging of electron emissions by cathodes. At that moment, such research appears to be of immediate relevance to the oscilloscope and television industry, but in the longer run it turns out that there is not that much to say about it. Brüche and his colleagues do give credit to the idea to use electron microscopes to obtain very high magnifications—reason enough even to come up with the word Übermikroskop—but this remains constrained to additional lip service, since they believe that fundamental problems of electron optics have to be overcome before high magnifications are feasible. They raise at least five concerns of which the foreseen problems with optical errors were certainly the most serious ones.

Ruska and von Borries, at the other hand, do not concern themselves at all with theoretical problems, but simply continue to travel the road on which Ruska already started under the guidance of Knoll. In 1931, Knoll and Ruska came up with the first two-step transmission images, the idea to encase the electron lens in order to reduce the focal length, the term "objective lens", a discussion of chromatic aberration, the suggestion to obtain very high magnifications of very small objects and the insight that a very high intensity of the electron beam would be needed. In 1932 and 1933 this record is expanded by von Borries and Ruska with patents on the polepiece lens, the intermediate screen and a cathode with high intensity, and with the first proper estimation of the resolution of the electron microscope, the construction of the first magnetic transmission electron microscope that was capable of surpassing the resolution limit of the light microscope, the first images to prove this, the first object chamber, the first realisation that elastic scattering would play a fundamental role in image formation, and finally with the term "ultramicroscopical structures", a call on physicians and biologists to join in, plus the suggestion to use metal dyes to stain biological objects.

From a retrospective point of view the situation has already become very clear, of course. Any electron microscopist who reads this, will have concluded by now that Knoll, Ruska and von Borries were the people who actually developed the instrument that we know nowadays as the classic transmission electron microscope. There might be some discussion about how much weight should be given to individual contributions, but in fact it is also rather clear that Ruska is the man who takes the biggest share, for the plain reason that he is the only one who was around continuously, from the first verification of the Busch lens theory in 1929 until the construction of the first supermicroscope by the end of 1933.

In the early 1930s, however, this perspective did not exist. First of all, the concept of the electron microscope proves to be very ill-defined at that time. It turns out that anything could be called electron microscope as soon as it was able to produce the slightest magnification by means of cathode rays. This is proven by its diverse use in very respectable scientific journals like *Annalen der Physik* and *Die Naturwissenschaften*, which shows that the term was accepted in its wide variety of meanings by science editors.

In fact, what we see on closer examination appears to be the emergence of two 'electron microscopical schools' by the end of 1933: the magnetic transmission school versus the electric emission school. The magnetic transmission school is represented by Ernst Ruska and Bodo von Borries who are focused on a type of instrument that is meant to surpass the resolution limit of light microscopy and with which transmission images can be made. From 1934 onwards, they will receive important support from Ladislaus Marton in Belgium. Reinhold Rüdenberg, Léo Szilárd and the American Paul Anderson could also be counted as members of this school, although they had not been actively involved in the development of a transmission electron microscope.

The major promoter of the electric emission school is without any doubt Ernst Brüche. In 1933 Vladimir Zworykin joins him in the USA, and also Max Knoll and Fritz Houtermans become closely associated with the emission school, despite the fact that they used a two-stage magnetic instrument to obtain emission images. Considering the fact that Knoll had moved to Telefunken in April 1932, Brüche worked for AEG and Zworykin for RCA, we also see that emission microscopy had become the domain of the electrotechnical industry, while magnetic transmission microscopy was left to others. The rationale for this has been explained already: an electrical lens is cheaper to manufacture, because of its simpler design, and at the same time a better understanding of cathode emissions has immediate relevance to the commercial production of oscilloscopes and televisions. In contrast to this, the electron optics of magnetic lenses had already proven to be complex and the prospect of making high resolution transmission images was rather doubtful, since you will have to apply a very intense electron beam, which would probably destroy the object. These misgivings had not yet been expressed very explicitly, but at the same time it is illustrative that the expressed doubts were not balanced by any attempts to start the development of an electrostatic instrument with a very high resolution. This sceptical attitude of the industry becomes even more apparent from the mere fact that both Ruska and von Borries were forced to find employment outside their newly established research field—at that stage commercial enterprises were not going to support the development of the transmission electron microscope. After 12 December 1933, Marton was the only one left who was actively involved with the development of the transmission electron microscope.

When it comes to the recurrent question about the motives of those who were involved in the development of the electron microscope, the answer is therefore very clear as far as electrostatic microscopy is concerned. The involvement of Brüche, Zworykin and others originated primarily from business interests. For the members of the transmission school, however, the matter is completely different. It appears that they were motivated above all by the archetypical, probably naïve and romantic idea of inventing a new technology that will change the future of science—at least for Ruska and Marton this seems to be the case. This last conclusion may be based on rather meagre evidence, but nevertheless it should be regarded as a fact that both men had started to focus on biological applications of the transmission electron microscope in early 1934.

From the perspective of the history and philosophy of science, the emergence of two electron microscope schools is probably the most interesting development so far. The moment the words "electron microscopy" started to appear in print, something like a discipline had sprouted, or at the very least it had been the intention of the men who used these words to raise that suggestion. Immediately attempts were made to define what an electron microscope is, in order to define the field of electron microscopy itself. We see this happen in a paper with the quite bold title "The Electron Microscope," which was published by the High Tension Lab (Knoll & Ruska, 1932a) and simultaneously in another paper with the just as outspoken title "Electron Optics and Electron Microscope" which was produced by the AEG Research Institute (Brüche & Johannson, 1932c). Their definitions left ample room for interpretation, however, and so the most successful player in the field will have influenced to a great extent which interpretation prevailed. It must be assumed that the very tangible results that were achieved with emission microscopy made much more impression at the time, if only for the reason that Ruska and colleagues had little to show for in 1933. Probably, the most impressive experimental result that Ruska had obtained was the image of the three-dimensional looking shadow of a metal extrusion, but rather ironically, this image was created with reflected electrons, and so you could see this work as an alternative form of emission studies as well (Ruska, 1933a). There is little reason, therefore, to believe that our contemporary definition of electron microscopy had any meaning to the vast majority of scientists and engineers in the early 1930s.

The emergence of different schools at the start of a new discipline is a theme that is extensively discussed by Thomas Kuhn in "The Structure of Scientific Revolutions" (Kuhn, 1962).

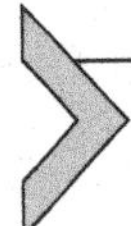

APPENDIX A. SIEMENS PATENT LIST

Group I, *German patents*

1. Reinhold Rüdenberg, German patent 889 660, Anordnung zur Beeinflussung des Verlaufs von Elektronenstrahlen durch elektrisch geladene Feldblenden, priority date of 31 May 1931, published 14 September 1953.
2. Reinhold Rüdenberg, German patent 895 635, Anordnung zur vergrößerten Abbildung von Gegenständen mittels Elektronenstrahlen und mittels den Gang der Elektronenstrahlen beeinflussender elektrostatischer oder elektromagnetischer Felder, priority date of 31 May 1931, published 5 November 1953.
3. Reinhold Rüdenberg, German patent 906 737, Anordnung zum vergrößerten Abbildung von Gegenständen mittels Elektronenstrahlen, priority date of 31 May 1931, published 18 March 1954.
4. Reinhold Rüdenberg, German patent 911 996, Einrichtung zum Abbilden von Gegenständen, priority date of 28 June 1931, published 24 May 1954.
5. Reinhold Rüdenberg, German patent 915 253, Anordnung zum Beeinflussen des Characters von Elektronenstrahlen durch elektrostatisch aufgeladene Doppelblenden, priority date of 13 August 1932, published 19 July 1954.
6. Reinhold Rüdenberg, German patent 916 838, Einrichtung zum vergrößerten Abbilden von Gegenständen durch Elektronenstrahlen, priority date of 27 June 1931, published 19 August 1954.
7. Reinhold Rüdenberg, German patent 916 839, Einrichtung zum vergrößerten Abbilden von Gegenständen durch Elektronenstrahlen, priority date of 31 March 1932, published 19 August 1954.
8. Reinhold Rüdenberg, German patent 916 841, Einrichtung zum vergrößerten Abbilden von Gegenständen durch Elektronenstrahlen

(Elektronenmikroskop), priority date of 31 March 1932, published 19 August 1954.

Group II, *non-German patents*

9. Reinhold Rüdenberg, Austrian patent 137,611, Einrichtung zum Abbilden von Gegenständen, priority date of 30 May, 26 and 27 June 1931 and 30 March 1932, published 25 May 1934.

10. Reinhold Rüdenberg, French patent 737,816, Disposition pour influencer la nature de rayons électroniques, priority date of 30 May 1931, published 16 December 1932.

11. Reinhold Rüdenberg, French patent 737,716, Dispositif pour obtenir des images d'objets, priority date of 30 May, 26 and 27 June 1931 and 30 March 1932, published 15 December 1932.

12. Reinhold Rüdenberg, British patent 402,781, Improvements in or relating to cathode-ray tubes, priority date of 30 May 1931, published 30 November 1933.

13. Reinhold Rüdenberg, Dutch patent 43263, Inrichting voor het vergroot afbeelden met behulp van een electronenmicroscoop met electrostatische lenzen, priority date of 30 May 1931, published 15 June 1938.

14. Reinhold Rüdenberg, Swiss patent 165,549, Einrichtung zum Abbilden von Gegenständen, priority date of 30 May, 26 and 27 June 1931 and 30 March 1932, published 2 April 1934.

15. Reinhold Rüdenberg, U.S. Patent 2,058,914, Apparatus for producing images of objects, priority date of 30 May 1931, granted on 27 October 1936.

16. Reinhold Rüdenberg, U.S. Patent 2,070,319, Apparatus for influencing the character of electron rays, priority date of 30 May 1931, granted on 9 February 1937.

Group III, *patents applied for by Swinne and Lübke*

17. Richard Swinne, German patent 915 843, Einrichtung zum Abbilden von Gegenständen, priority date of 18 June 1932, published 29 July 1954.

18. Richard Swinne, German patent 916 840, Einrichtung zum vergrößerten Abbilden von Gegenständen in zwei Stufen, priority date of 18 June 1932, published 19 August 1954.

19. Ernst Lübcke, German patent 659 359, Gittergesteuerte Entladungsröhre mit einem Lichtbogen als Elektronenquelle, priority date of 1 August 1933, published 2 May 1938.

Group IV, related patents

20. Siemens-Schuckertwerke AG, German Patent 754 259, Anordnung zur Steuerung der Intensität der durch eine Lochblende hindurchtretenden Elektronenstrahlen, priority date of 31 May 1931, granted on 31 August 1944, never published.

21. Siemens-Schuckertwerke AG, patent 758 391, Anordnung zur Beeinflussung des Verlaufs von Elektronenstrahlen durch platten- oder ringförmige elektrisch geladene Feldblenden, priority date of 31 May 1931, date on which it was granted unknown, never published.

22. Siemens & Halske AG, German patent 909 156, Einrichtung zur Erzeugung von Gefügebildern mit Elektronenstrahlen, priority date of 15 May 1932, published 11 March 1954.

APPENDIX B. POSTSCRIPT: A TRIBUTE TO JOHN VAN GORKOM BY DIRK VAN DELFT AND TON VAN HELVOORT

"Es war eine Manie"

In the course of his studies, the biologist John van Gorkom became fascinated by the history of the electron microscope and in particular its inventor, Ernst Ruska. Sadly and unexpectedly, van Gorkom died in 2013, leaving his research on this topic at an advanced but incomplete stage. In the foregoing pages, we offer excerpts from his manuscript.

John van Gorkom was born on February 19, 1960 in Penrith, Australia, just fifty kilometers west of Sydney, where his parents had emigrated from the Netherlands the previous year. The emigration down under did not prove to be a success for, homesick, the van Gorkoms returned to Europe within five years.

Back in the Netherlands, the family settled down in Delft, where van Gorkom attended secondary school at the Stedelijke Scholengemeenschap 'Hugo Grotius' (now *Grotius College*). Among his classmates there was his future wife Caroline Le Poole, daughter of the Dutch electron microscopist and inventor Jan B. Le Poole. After completing school, both young people went on to study biology at Utrecht, where they fell in love and became a couple. Thus, John was to become "infected" with the electron microscope bug, early in life, through his future father-in-law.

In deference to the wishes of Caroline's family, the young couple got married in 1981. John, who had been greatly impressed by Berlin while

Wedding photograph of John van Gorkom and Caroline Le Poole, 7 August 1981.

on a visit there in 1978 with a classmate Bart de Haas, suggested it as a destination for their honeymoon. When his father-in-law heard about this choice, he called upon his old friend, Ernst Ruska to make arrangements for the honeymooners to visit.

John and Caroline arrived in Berlin in the midst of the celebrations, admixed with regrets, of the twentieth anniversary of the Berlin Wall. As

Irmela and Ernst Ruska, accompanied by John and Caroline, on a terrace near *Die Brücke der Einheit* in Berlin.

arranged, they visited Ruska, by then 74 years old, and his wife Irmela. He proved an amiable host who regaled them with trips down memory lane about his life and work. He also arranged for a visit to the institute in East Berlin, where his brother Helmut had produced the first electron microscopic pictures of viruses in the late 1930s. According to Caroline, John's fascination with the electron microscope began during this visit, although it was not manifested immediately. Indeed about two months after their return to Utrecht, they received an autographed copy of Ernst Ruska's *Die frühe Entwicklung der Elektronenlinsen und der Elektronenmikroskopie* in the mail, but it disappeared unread into their closet.

In 1984, John and Caroline made a second visit to Berlin, where they proudly showed off their 7-month old daughter Emy to the Ruskas, *Onkel* Ernst and *Aunt* Irmer. Although named after her great-grandmother, Jan Le Poole's mother, John was pleased to point out that she was also named for the electron microscope (abbreviated EM) that had captured his interest. It was around this time that he began to wonder why Ruska had not received a Nobel Prize for his work. Back in Netherlands he started to pursue this line of thought more intentionally, and in the summer of 1985 returned yet again to Berlin with his friend de Haas in order to conduct in-depth interviews with Ruska. The outcome was a series of three sessions that together amounted to 10 hours of conversation. After their return to the Netherlands he also recorded an interview with Le Poole (these recordings are now archived at the Rijksmuseum Boerhaave, Leiden).

Meanwhile, unfortunately John's personal life had taken turn for the worse. His marriage began to deteriorate and by 1986 ended in a divorce. He also suffered very slow progress in the laboratory, due to which he decided to abandon this research. Although he had come up with the idea of redirecting his attention and writing about the history of electron microscopy for his graduate thesis, he made only little progress on that front due to his personal and professional problems.

Later in 1986 however, against all expectations, Ruska was named as one of the recipients of the Nobel Prize in Physics for inventing the electron microscope. This announcement revived John's interest in Ruska and raised several questions. Why, for instance, did he have to wait 55 years to be recognized with the Nobel? How had he managed to sustain his work despite initial resistance from the very biologists who ultimately stood to gain the most from his invention and the lack of interest from commercial companies such as Siemens and Zeiss, who would later manufacture the instrument. Ruska's uniform response to all such questions was "Es war eine Manie" or in plain English, "It was a mania." This refrain fitted van Gorkom's project like a glove, and he used it as the title for his Master's thesis, which he completed in 1988, just around the time of Ruska's death.

Almost two decades later, John returned to the topic and in 2004 completed a book-length manuscript in English on the topic, titled *55 Years of Waiting—Ernst Ruska and the history of the electron microscope*. But he was unable to garner interest from English-language publishers or literary agents and the writings remained unpublished. Meanwhile however, he did not give up, and in 2010 succeeded in interesting Dirk van Delft, director of the Rijksmuseum Boerhaave and professor, by special appointment, of the "Material heritage of the natural sciences" at Leiden University. After a couple of interviews, the two men came to an arrangement for van Delft to supervise John van Gorkom's doctoral thesis on the development of the electron microscope.

Over the next couple of years John would work with such intensity and fervor that he could talk of little else and his family thought he was manic.

By 2013 he had produced a sound body of work—the first six chapters of his doctoral thesis, which focused mainly on the physics and engineering of the electron microscope. But he had not yet arrived at the topic that had been at the heart of his intention for the product, which was the history of the applications of the instrument to studying biological objects. To this end, in August 2013 he travelled to Berlin again with his old friend Bart de Haas. During that week, he visited the Ruska archives and consulted historians of science. Some weeks later toward the end of September, back at home in Amsterdam, while preparing for a new day, his heart stopped beating and he died at age 53.

John van Gorkom's project however, did not die with him. The archival material he had gathered, together with his manuscripts were transferred to his advisor, who then approached science writer and historian of biomedicine, Ton van Helvoort for help in bringing the work to print although in a more popular format. The successful outcome of this collaboration is their aptly named *Beelden zonder weerga* (in English: *Images unprecedented: The electron microscope from Ernst Ruska to Ben Feringa*), published this year by Prometheus Press in Amsterdam. Whereas this book is for a wider audience, physicists and historians alike will be pleased to read in these pages, the central two chapters of John's own words on the subject that so engaged him. In the wake of the award of the 2017 Nobel Prize in Chemistry to the inventors of cryo-electron microscopy, now is an especially apropos time to read about the history of the field that in its many guises—transmission, scanning, and cryo-electron microscopy—has revealed, and continues to reveal, information about life through images unprecedented.

REFERENCES

Brüche, E. (1930). Strahlen langsamer Elektronen und ihre technische Anwendung. In W. Petersen (Ed.), *Forschung und Technik* (pp. 23–46). Berlin: Springer.

Brüche, E. (1931/32). Geometrische Elektronenoptik. *Jahrbuch des Forschungs-Instituts der Allgemeinen Elektricitäts-Gesellschaft*, *3*, 111–124 (published May or June 1933).

Brüche, E. (1932a). Elektronenmikroskop. *Naturwissenschaften*, *20*, 49.

Brüche, E. (1932b). Die Geometrie des Beschleunigungsfeldes in ihrer Bedeutung für den gaskonzentrierten Elektronenstrahl. *Zeitschrift für Physik*, *78*, 26–42.

Brüche, E. (1932c). Biegsame Elektronenstrahlen. *Zeitschrift für Physik*, *78*, 177–195.

Brüche, E. (1932d). Geometrische Elektronenoptik. *Forschungen und Fortschritte*, *8*, 316.

Brüche, E. (1933). Elektronenmikroskopische Abbildung mit lichtelektrischen Elektronen. *Zeitschrift für Physik*, *86*, 448–450.

Brüche, E. (1940). 10 Jahre Elektronenmikroskopie bei der AEG. *AEG Mittelungen*, 302–307.

Brüche, E., & Johannson, H. (1932a). Kinematographische Elektronenmikroskopie von Oxydkathoden. *Annalen der Physik*, *407*, 145–166.

Brüche, E., & Johannson, H. (1932b). Einige neue Kathodenuntersuchungen mit dem elektrischen Elektronenmikroskop. *Physikalische Zeitschrift*, *33*, 898–899.

Brüche, E., & Johannson, H. (1932c). Elektronenoptik und Elektronenmikroskop. *Naturwissenschaften*, *20*, 353–358.

Busch, H. (1926). Berechnung der Bahn van Kathodenstrahlen im axialsymmetrischen elektromagnetischen Felde. *Annalen der Physik*, *386*, 974–993.

Busch, H. (1927). Über die Wirkungsweise der Konzentrierungsspule bei der Braunschen Röhre. *Archiv für Elektrotechnik*, *18*, 583–594.

Davisson, C. J., & Calbick, C. J. (1931). Electron lenses. *Physical Review*, *38*, 585. A year later an adjustment was published in Davisson and Calbick (1932).

Davisson, C. J., & Calbick, C. J. (1932). Electron lenses. *Physical Review*, *42*, 580.

de Broglie, L. (1946). Mécanique ondulatoire et optique électronique. In L. de Broglie (Ed.), *L'optique electronique* (pp. 1–17). Paris: Editions de la Revue d'Optique.

Feld, B. T. (2008). Szilárd, Leo. In *Complete dictionary of scientific biography: Vol. 13* (pp. 226–228). Detroit: Charles Scribner's Sons. Gale Virtual Reference Library, http://go.galegroup.com/ps/. (Accessed 9 September 2012).

Flaningam, M. L. (1945). International co-operation and control in the electrical industry, the General Electric Company and Germany, 1919–1944. *American Journal of Economics and Sociology*, *5*(1), 7–25.

Fournier, M. (2009). Electron microscopy in Second World War Delft. In A. Maas, & H. Hooijmaijers (Eds.), *Scientific research in World War II, what scientists did in the war* (pp. 77–95). Routledge.

Frank, T. (2008). Szilárd, Leo. In *Complete dictionary of scientific biography: Vol. 24* (pp. 573–576). Detroit: Charles Scribner's Sons. Gale Virtual Reference Library, http://go.galegroup.com/ps/. (Accessed 9 September 2012).

Freundlich, M. M. (1994). The history of the development of the first high-resolution electron microscope. *Microscopy Society of America Bulletin*, *24*, 405–415.

Fritz, R. (1936). Le microscope électronique. *Revue Generale des Sciences Pures et Appliquées*, *47*, 338–342.

Gabor, D. (1942). Electron optics. *Electronic Engineering*, *15*, 295–299.

Gabor, D. (1957). Die Entwicklungsgeschichte des Elektronenmikroskops. *Elektrotechnische Zeitschrift-A*, *78*, 522–530.

Gabor, D. (1974). The history of the electron microscope, from ideas to achievements. In *Electron microscopy 1974, Eighth international congress on electron microscopy: Vol. 1* (pp. 6–12). Canberra: Australian Academy of Science.

Glaser, W. (1933a). Theorie des Elektronenmikroskopes. *Zeitschrift für Physik*, *83*, 104–122.

Glaser, W. (1933b). Über geometrisch-optische Abbildung durch Elektronenstrahlen. *Zeitschrift für Physik*, *80*, 451–464.

Glaser, W. (1933c). Zur geometrischen Elektronenoptik des axialsymmetrischen elektromagnetischen Feldes. *Zeitschrift für Physik*, *81*, 647–686.

Grünewald, H. (1976). Zur Entstehungsgeschichte des Elektronenmikroskops mit elektromagnetischen Linsen. *Technikgeschichte*, *43*, 213–222 (p. 213).

Hawkes, P. W. (1985). Complementary accounts of the history of electron microscopy. In P. W. Hawkes (Ed.), *Advances in electronics and electron physics: Suppl. 16. The beginnings of electron microscopy* (pp. 589–618). Orlando: Academic Press.

Howie, A. (1982). A harnessing of electrons. *Science*, *216*, 878–880.

Isherwood, C. (1935). Mr Norris changes trains. Fragments from first page of Chapter 16 in Isherwood, C. (1999). The Berlin novels. London: Vintage.

Johannson, H. (1933). Über das Immersionsobjektiv der geometrischen Elektronenoptik. *Annalen der Physik, 410*, 385–413.

Johannson, H., & Scherzer, O. (1933). Über die elektrische Elektronensammellinse. *Zeitschrift für Physik, 80*, 183–192.

Keith, S. T. (2008). Gabor, Dennis. In *Complete dictionary of scientific biography: Vol. 17* (pp. 324–328). Detroit: Charles Scribner's Sons. Gale Virtual Reference Library, http://go.galegroup.com/ps/. (Accessed 9 September 2012).

Knoll, M. (1929). Vorrichtung zur Konzentrierung des Elektronenstrahls eines Kathodenstrahloszillographen. German patent 690809, filed on 10 November 1929, granted on 11 April 1940. Lin argues that Knoll's patent merely describes a concentrating device, since Knoll also proposes an arrangement that is composed of five sheets with opposite charges, as an alternative for the arrangement with three sheets.

Knoll, M. (1931). Berechnungsgrundlagen und neuere Ausführungsformen des Kathodenstrahloszillographen. Manuscript of a lecture in the Cranz-Colloquium of the Technische Hochschule Berlin on 4 June 1931, 26 pages. Abstract published in Ruska (1979), Anhang B, p. 116. Also published in Ruska (1984), pp. 530 and 531, and Ruska (1986).

Knoll, M., Houtermans, F. G., & Schulze, W. (1932a). Über geometrisch-optische Abbildung von Glühkathoden durch Elektronenstrahlen mit Hilfe von Magnetfeldern (Elektronenmikroskop). *Verhandlungen der deutschen physikalischen Gesellschaft, 13*, 23–24.

Knoll, M., Houtermans, F. G., & Schulze, W. (1932b). Untersuchung der Emissionsverteilung an Glühkathoden mit dem magnetischen Elektronenmikroskop. *Zeitschrift für Physik, 78*, 340–362.

Knoll, M., Houtermans, F., & Schulze, W. (1932c). Elektronenmikroskop, bei dem Elektronen aussendende Substanzen in vergrößertem Maßstabe abgebildet werden. German patent 679330, priority date of 17 March 1932, published on 13 July 1939.

Knoll, M., & Ruska, E. (1932a). Das Elektronenmikroskop. *Zeitschrift für Physik, 78*, 318–339.

Knoll, M., & Ruska, E. (1932b). Beitrag zur geometrischen Elektronenoptik (I, II). *Annalen der Physik, 404*, 607–640, 641–661.

Kraus, P. (1938). Julius Ruska. *Osiris, 5*, 4–40.

Kuhn, T. S. (1962). *The structure of scientific revolutions*. Chicago & London: University of Chicago Press (3rd ed., 1996).

Lambert, L., & Mulvey, T. (1996). Ernst Ruska (1906–1988), designer extraordinaire of the electron microscope: A memoir. *Advances in Imaging and Electron Physics, 95*, 3–62.

Landrock, K. (2003). Friederich Georg Houtermans (1903–1966) – Ein bedeutender Physiker des 20 Jahrhunderts. *Naturwissenschaftliche Rundschau, 56*, 187–199. Austrian Academy of Sciences, Fritz Georg Houtermans (1903–1966). http://www.oeaw.ac.at/shared/news/2003/inf_houtermans.html. (Accessed 20 October 2012).

Lenard, P. (1918). *Quantitatives über Kathodenstrahlen aller Geschwindigkeiten*. Heidelberg: C. Winters.

Le Poole, J. B. (1985). Early electron microscopy in the Netherlands. In P. W. Hawkes (Ed.), *Advances in electronics and electron physics: Suppl. 16. The beginnings of electron microscopy* (pp. 387–416). Orlando: Academic Press.

Lin, Q. (1995). *Zur Frühgeschichte des Elektronenmikroskops*. Stuttgart: Verlag für Geschichte der Naturwissenschaften und der Technik.

Marton, L. (1968). *Early history of electron microscopy*. San Francisco: San Francisco Press.

Marton, L., & Nuyens, M. (1933). Meetkundige optiek der electronen. *Wis- en Natuurkundig Tijdschrift, 6*, 159–170.

Matthias, A., von Borries, B., & Ruska, E. (1933). Eine neue Form des Strom-Messsystems am Kathodenstrahloszillographen. *Zeitschrift für Physik, 85*, 336–352.

Müller, F. (2009). The birth of a modern instrument and its development during World War II: Electron microscopy in Germany from the 1930s to 1945. In A. Maas, & H. Hooijmaijers (Eds.), *Scientific research in World War II, what scientists did in the war* (pp. 121–146). Abingdon: Routledge.

Mulvey, T. (1962). Origins and historical development of the electron microscope. *British Journal of Applied Physics, 13*, 197–207.

Mulvey, T. (1973). Forty years of electron microscopy. *Physics Bulletin, 24*, 147–154 (p. 149). The suggestion is repeated in Lambert and Mulvey (1996).

Perry, R. P. (2005). Thomas Foxen Anderson. In *Biographical memoirs: Vol. 87* (pp. 51–72). Washington: National Academic Press.

Petersen, W. (Ed.). (1930). *Forschung und Technik*. Berlin: Springer.

Rajchman, J. (2006). Vladimir Kosma Zworykin 1889–1982. In *Biographical memoirs: Vol. 88* (pp. 1–21). Washington: National Academic Press.

Ramsauer, C. (1941). *Zehn Jahre Elektronenmikroskopie: Ein Selbstbericht des AEG Forschungs-Instituts*. Berlin: Springer-Verlag.

Ramsauer, C. (1949). *Physik – Technik – Pädagogik, Erfahrungen und Erinnerungen*. Karlsruhe: G. Braun.

Rasmussen, N. (1997). *Picture control, the electron microscope and the transformation of biology in America, 1940–1960*. Stanford: Stanford University Press.

Reisner, J. H. (1989). An early history of the electron microscope. *Advances in Electronics and Electron Physics, 73*, 133–231.

Rogowski, W., & Graupner, H. (1932). Zur Erklärung des gaskonzentrierten Elektronenstrahls. *Archiv für Elektrotechnik, 26*, 807–810.

Royal Academy of Sciences (1986). The Nobel Prize in Physics 1986 press release, 15 October, second paragraph. www.nobelprize.org.

Rüdenberg, R. (1932). Elektronenmikroskop. *Naturwissenschaften, 20*, 522.

Rüdenberg, R. (1943). The early history of the electron microscope. *Journal of Applied Physics, 14*, 434–436.

Rüdenberg, R. (2010). Origin and background of the invention of the electron microscope. In P. Hawkes (Series Ed.), *Advances in imaging and electron physics: Vol. 160* (pp. 171–205). Academic Press. The text is the reproduction of the latest draft of a memoir from 1959, as explained on the first page of the article.

Rudenberg, H. G., & Rudenberg, P. G. (2010). Origin and background of the invention of the electron microscope: Commentary and expanded notes on memoir of Reinhold Rüdenberg. In P. Hawkes (Series Ed.), *Advances in imaging and electron physics: Vol. 160* (pp. 207–286). Academic Press (p. 236). Gunther dropped the family name's umlaut as he was a US citizen. He died in January 2009 according to Peter Hawkes in his preface to the above-mentioned volume in which the Rüdenberg contribution appeared. His son Paul finished and submitted it.

Rupp, E. (1928). Über Elektronenbeugung an einem geritzten Gitter. *Zeitschrift für Physik, 52*, 8–15.

Rupp, E. (1931). Über die Gültigkeit der de Broglieschen Beziehung für sehr schnelle Elektronen (220 kV). *Annalen der Physik, 401*, 458–464.

Ruska, E. (1929). *Über eine Berechnungsmethode des Kathodenstrahloszillographen auf Grund der experimentell gefundenen Abhängigkeit des Schreibfleckdurchmessers von der Stellung der Konzentrierspule* (Studienarbeit). Berlin: Technische Hochschule Berlin, Lehrstuhl für Hochspannungstechnik. Retrieved from http://ernstruska.digilibrary.de.

Ruska, E. (1930). Untersuchung elektrostatischer Sammelvorrichtungen als Ersatz der magnetischen Konzentrierspulen beim Kathodenstrahloszillographen (Diplomarbeit). Technische Hochschule Berlin, Lehrstuhl für Hochspannungstechnik. Retrieved from http://ernstruska.digilibrary.de.

Ruska, J. (1932). Weltbild und Naturforschung im Wandel der Zeiten. *Fortschritte der Medizin, 50*, 88–91.

Ruska, E. (1933a). Die elektronenmikroskopische Abbildung elektronenbestrahlter Oberflächen. *Zeitschrift für Physik, 83*, 492–497.

Ruska, E. (1933b). Zur Fokussierbarkeit von Kathodenstrahlbündeln grosser Ausgangsquerschnitte. *Zeitschrift für Physik, 83*, 684–697.

Ruska, E. (1934a). Über Fortschritte im Bau und in der Leistung des magnetischen Elektronenmikroskops. *Zeitschrift für Physik, 87*, 580–602.

Ruska, E. (1934b). Über ein magnetisches Objektiv für das Elektronenmikroskop (Dissertation, Technische Hochschule Berlin). *Zeitschrift für Physik, 89*, 90–128.

Ruska, E. (1934c). Das Elektronenmikroskop als Übermikroskop. *Forschungen und Fortschritte, 10*, 8 (submission date mentioned in Qing, 1995).

Ruska, E. (1957). 25 Jahre Elektronenmikroskopie. *Elektrotechnische Zeitschrift-A, 78*, 531–543.

Ruska, E. (1970a). Erinnerungen an die Anfänge der Elektronenmikroskopie. In *Festschrift anlässlich der Verleihung des Paul-Ehrlich- und Ludwig-Darmstaedter-Preises 1970* (pp. 19–34). Stuttgart: Gustav-Fischer-Verlag.

Ruska, H. (1970b). Bisherige Erfolge und künftige Ziele der Elektronenmikroskopie. In *Festschrift anlässlich der Verleihung des Paul-Ehrlich- und Ludwig-Darmstaedter-Preises 1970* (pp. 35–48). Stuttgart: Gustav-Fischer-Verlag.

Ruska, E. (1974). Zur Vor- und Frühgeschichte des Elektronenmikroskops. In *Electron microscopy 1974, Eighth international congress on electron microscopy: Vol. 1*. Canberra: Australian Academy of Science.

Ruska, E. (1979). Die frühe Entwicklung der Elektronenlinsen und der Elektronenmikroskopie. *Acta Historica Leopoldina, 12*, 1–136.

Ruska, E. (1980). *The early development of electron lenses and electron microscopy*. Stuttgart: Hirzel.

Ruska, E. (1984). Die Entstehung des Elektronenmikroskops (Zusammenhang zwischen Realisierung und erster Patentanmeldung, Dokumente einer Erfindung). *Archiv der Geschichte der Naturwissenschaften, 11*, 525–551.

Ruska, E. (1986). The emergence of the electron microscope (Connection between realization and first patent application, Documents of an invention). *Journal of Ultrastructure and Molecular Structure Research, 95*, 3–28.

Ruska, E. (1987). The development of the electron microscope and of electron microscopy, Nobel lecture, December 8, 1986. *Reviews of Modern Physics, 59*, 627–638.

Ruska, E., & Knoll, M. (1931). Die magnetische Sammelspule für schnelle Elektronenstrahlen. *Zeitschrift für technische Physik, 12*, 389–399, 448.

Scherzer, O. (1933). Zur Theorie der elektronenoptischen Linsenfehler. *Zeitschrift für Physik, 80*, 193–202.

Störmer, C. (1929). Kurzwellenechos, die mehrere Sekunden nach dem Hauptsignal eintreffen, und wie sie sich aus der Theorie des Polarlichtes erklären lassen. *Naturwissenschaften, 17*, 643–651.

Süsskind, C. (1985). L.L. Marton, 1901–1979. In P. W. Hawkes (Ed.), *Advances in electronics and electron physics: Suppl. 16. The beginnings of electron microscopy* (pp. 501–523). Orlando: Academic Press.

Szilárd, L. (1931). Mikroskop. German patent 965 522, priority date of 4 July 1931, published 29 May 1957.

Thomson, G. P. (1931). Electron optics. *Nature, 129*, 81–83.

Trillat, J.-J. (1962). Some personal reminiscences. In P. P. Ewald (Ed.), *Fifty years of X-ray diffraction* (pp. 662–666). Utrecht: Oosthoek's Uitgeversmaatschappij.

Trillat, J.-J. (1981). The start of electron diffraction in France: My recollections. In P. Goodman (Ed.), *Fifty years of electron diffraction* (pp. 77–79). Dordrecht: D. Reidel.

van Gorkom, J., & de Haas, B. (1985). Interview with Ernst Ruska in Berlin on 25 June, 27 June and 3 July 1985.

von Borries, B., & Ruska, E. (1932a). Das kurze Raumladungsfeld einer Hilfsentladung als Sammellinse für Kathodenstrahlen. *Zeitschrift für Physik, 76*, 649–654.

von Borries, B., & Ruska, E. (1932b). Magnetische Sammellinse kurzer Feldlänge. German patent 680284, priority date of 17 March 1932, published on 3 August 1939.

von Borries, B., & Ruska, E. (1933a). Zur Erklärung des gaskonzentrierten Elektronenstrahls, Bemerkung zu der Arbeit von Rogowski und Graupner. *Archiv für Elektrotechnik, 27*, 227. A response to Rogowski and Graupner (1932).

von Borries, B., & Ruska, E. (1933b). Die Abbildung durchstrahlter Folien im Elektronenmikroskop. *Zeitschrift für Physik, 83*, 187–193.

von Borries, H. (1991). Bodo von Borries. *Advances in Electronics and Electron Physics, 81*, 127–176.

von Weiher, S. (2008). Rüdenberg, Reinhold. In *Complete dictionary of scientific biography: Vol. 11* (pp. 588–589). Detroit: Charles Scribner's Sons. Gale Virtual Reference Library, http://go.galegroup.com/ps/. (Accessed 9 September 2012).

Watson, W. (1752). An account of the phaenomena of electricity in vacuo, with some observations thereupon. *Philosophical Transactions of the Royal Society of London, 47*, 362–376.

Wiechert, E. (1899). Experimentelle Untersuchungen über die Geschwindigkeit und die magnetische Ablenkbarkeit der Kathodenstrahlen. *Annalen der Physik und Chemie, 305*, 739–766.

Wolpers, C. (1991). Electron microscopy in Berlin 1928–1945. *Advances in Electronics and Electron Physics, 81*, 211–229.

Zworykin, V. K. (1933). On electron optics. *Journal of the Franklin Institute, 215*, 535–555.

CHAPTER TWO

Electron Optics of Low-Voltage Electron Beam Testing and Inspection. Part I: Simulation Tools*

Erich Plies

Forschungslaboratorien, Siemens AG, München, Germany
Present address: Institute for Applied Physics, University of Tübingen, Tübingen, Germany.
e-mail address: erich.plies@uni-tuebingen.de

Contents

List of Abbreviations

BEM Boundary element method
CD Critical dimension

* Reprinted from Advances in Optical and Electron Microscopy, vol. 13, 1994, 123–242.

ISSN 1076-5670
https://doi.org/10.1016/bs.aiep.2017.12.001

CDM Surface charge density method
CSM Charge simulation method
DRAM Dynamic random access memory
DUT Device under test
e-beam Electron beam
EBIC Electron beam induced current analysis
EBIS Electron beam inspection system
EBT Electron beam tester
EDX Energy dispersive X-ray analysis
FDM Finite difference method
FE Field emission
FEM Finite element method
FWHM Full width at half maximum
HPCD Hamming's modified predictor–corrector method (HPC) in difference form
IC Integrated circuit
IEM Integral equation method
LFE Local field effect
LSFM Least-squares fit method
MOSFET Metal oxide semiconductor field effect transistor
PC Personal computer
PDE Partial differential equation
PE Primary electron
rms Root mean square
SCWIM Spherical coordinates with increasing mesh
SE Secondary electron
SEM Scanning electron microscope
SLOR Successive line over-relaxation method
SOEM Second-order electrode method
SOR Successive over-relaxation method
SWEM Spherical coordinates with exponentially increasing mesh
VAIL Variable axis immersion lens

1. INTRODUCTION

The first references to the basic idea of the scanning electron microscope (SEM) can be found in the publications of Knoll (1935) and von Ardenne (1938a, 1938b). The development of the first commercial SEMs for directly imaging surfaces was not completed until the 1960s. Today, there is a great variety of specialized electron optical devices which have been developed from the SEM. These devices include the electron beam (e-beam) tester (EBT), the e-beam inspection system (EBIS) and other equipment similar to the EBIS which are all used in the semiconductor industry. The electron optical column is the main component of this type of equipment. Apart from a few inspection applications, this equipment is operated in the low-voltage regime of about 1 kV. The first specially developed low-voltage optics are now available, while the majority of SEMs and similar equipment still manage with a classical medium-voltage column, i.e. two magnetic condenser lenses and one magnetic objective lens, operating them down to 1 keV, instead of beam energies between 5 and 40 keV.

The first early attempts to improve considerably the resolution of the SEM operating in the low-voltage regime by means of analyses based on electron optics are found in publications by Zworykin, Hillier, and Snyder (1942) and Pease (1967). In spite of this pioneering work, we are still at the start of theoretical approaches to the optimization of compound, low-voltage electron optics with, as one would expect, experimental work making even less progress. Low-voltage electron optics are relevant not only to semiconductor development but also to biology, for example. As biological specimens are usually very small, they can be positioned in the gap region of a magnetic lens (Nagatani, Saito, Sato, & Yamada, 1987). This automatically reduces aberrations and improves resolution. On the other hand, however, requirements in the field of biology are 2 nm for 1.5 keV electrons (Pawley, 1990), i.e. three times better than the highest known resolution needed for semiconductor development, at the same acceleration voltage. The semiconductor industry, however, wants to subject very large samples, wafers of up to 10 inches (25 cm) in diameter, to topographical analysis. Moreover, this has to be in the tilted state of the wafer, which requires a separation of a few millimeters from the last electrode or pole face. The electron optical challenge in the semiconductor industry can thus be considered to be equally important. The special electron optical requirements for biological applications will not be discussed further in this chapter.

A major reason for limiting the probe diameter in the low-voltage regime is the dominance of chromatic aberration. Preliminary calculations show that a resolution of 10 nm at a beam energy of 0.5 keV is likely to be obtained in spite of a working distance of 3–4 mm both by means of a retarding electrostatic probe-forming objective lens (Zach, 1989) and by means of a compound magnetic and electrostatic lens (Frosien, Plies, & Anger, 1989). It is probable that a further considerable increase in resolution at low electron energies of this kind could be obtained by correcting the chromatic aberration (Zach, 1989). Apart from resolution, i.e. the probe diameter, the probe current is another important characteristic for e-beam testing and electron optical inspection, if practicable measurement times and high throughput rates are to be obtained. In the low-voltage regime, however, the source brightness declines sharply, more sharply even than if it were directly proportional to the beam voltage as predicted by Langmuir's equation. This means that there is less beam current in a probe of a particular diameter or that a larger probe diameter is required to obtain a higher current. Also, an anomalous energy spread occurs at high beam currents in the low-voltage operating regime due to the Boersch effect (Boersch, 1954). This causes an additional increase in the probe diameter by the chromatic aberration of the objective lens. Low-voltage optics must therefore be minimized to take the Boersch effect into account.

In Part I, after two short introductory sections on e-beam testing and on electron optical inspection, the calculation of fields and trajectories, i.e. electron optical simulations will first be discussed. Although the special electron optical problems associated with e-beam testing and inspection are discussed in this somewhat more theoretical section, Part I is also largely concerned with other electron optical applications. Part II deals with the electron optical components and systems used for e-beam testing and inspection. This topic was already treated very briefly by Plies (1990a) but 183 references were compiled.

1.1 e-Beam Testing

e-Beam testing is a powerful tool for design and chip verification as well as for performing failure analysis of integrated circuits (ICs). e-Beam testing originated with voltage contrast imaging of devices under test (DUTs) in SEMs (Oatley & Everhart, 1957). The electron probe also makes it possible to perform waveform measurements at internal nodes and to image the logic operations of entire parts inside the DUT. As the interconnection width decreased, e-beam testing became increasingly common and has replaced test techniques using a mechanical tip (Plies & Otto, 1985). The major advantages of e-beam testing are the fineness of the electron probe which can be quickly positioned on submicrometer interconnections, and its non-loading and non-destructive applications at low primary electron (PE) energies.

Until the middle of the 1980s, modified SEMs were used to perform e-beam testing on ICs. The electron optical modifications to the SEM usually comprised a beam-blanking system between the anode and the first magnetic condenser lens, and a simple retarding-field spectrometer between the magnetic objective lens and the DUT. Today, dedicated EBTs are available. In these devices, the spectrometer is situated in or over the objective lens, and, because of the reduced working distance, an electron probe with a diameter of 0.1–0.3 μm and a beam current greater than 1 nA at a low beam voltage of 1 kV can be produced. The low energy of the PEs is necessary to avoid damage, charging and loading of the IC. Unlike those in the modified SEMs, the new spectrometers were also successful in improving the detection of secondary electrons (SEs), thus increasing the precision of voltage measurements and reducing sensitivity to the effects of the local fields of the IC. It is still desirable to reduce the local field effect (LFE) further, and to make additional electron optical improvements, in particular for voltage measurements on interconnections narrower than 0.7 μm. A further increase in the beam current to decrease measuring times, or to improve the signal-to-noise ratio of the detected SEs, would also be welcome.

The reviews of Argyo et al. (1985), Wolfgang (1986), Ura and Fujioka (1989), and Gopinath (1987, 1989) should be consulted for the various e-beam testing modes and methods, signal processing, and automatic control of the EBT and the test methodology. This article will describe only the low-voltage electron optics of the EBT from the standpoint of both PEs and SEs.

1.2 Low-Voltage Electron Optical Inspection

Although only the term "inspection" appears in the title of this section, both here and in the following sections we will also be dealing with metrology systems. This is because, as far as the electron optical column is concerned, differences are slight. This will become apparent at the appropriate point.

Classical optical microscopy has long been the standard tool for inspection and metrology of ICs. Improved optical techniques such as confocal-scanning laser microscopy (Wijnaendts van Resandt & Zapf, 1988), coherence probe microscopy (Davidson, Kaufman, Mazor, & Cohen, 1987) and holographic inspection (Billat, 1988) are still being used successfully today. However, because the minimal feature sizes of ICs are continuously decreasing, the number of SEM-based inspection and measurement devices used by the semiconductor industry is increasing sharply.

Electron microscopy at about 20 keV has established itself as the technology of choice for off-line process control and off-line quality assurance applications where cross-sectioning and/or conductive coating are permitted. In-line applications, such as detecting patterning defects, defect analysis, measurements of critical dimensions (CDs), e.g. line widths and spacings, and measurements of overlay accuracy, require low beam voltages (0.8–1.3 kV) to avoid radiation damage and charging of uncoated and non-conducting structures which are part of the wafer or on it, e.g. etched resist structures. CD measurement SEMs with a resolution of 15 nm at 1 keV are now commercially available (Ohtaka, Saito, Furuya, & Yamada, 1985; Shimada, Mimura, Sawaragi, Suzuki, & Aihara, 1986). Low-voltage equipment for defect detection has already been announced and there is a demand for low-voltage defect analysis instruments. The resolution requirements for defect analysis are the most stringent (6 nm at 1 keV). For defect detection, the probe requirements are also very exacting (a few tens of nanoamps) to ensure a high throughput. Photo and X-ray masks are usually inspected at higher acceleration voltages (3–20 kV) because of the materials used and the resulting contrast conditions. In-process examination of wafers—no matter whether this involves CD measurements, defect inspection or defect review—must always be performed at low voltages. We will discuss only the electron optics for these low-voltage optics.

In the same way that in-lens or through-lens analyzers have ousted post-lens analyzers in the field of e-beam testing, the use of through-lens detection of SEs or back-scattered electrons will win through in e-beam inspection applications. Even today, many SEMs use an Everhart–Thornley detector above the magnetic objective lens to detect SEs after they have spiralled up the magnetic field of the objective lens. However, this unilaterally arranged detector has too strong an effect on low-energy PE beams (deflection, astigmatism); there is also room for improvement in terms of collection efficiency. The detector arrangements must be treated in more depth at the electron optical design stage, for example by using an electrode and detector arrangement with rotational symmetry as suggested by Zach and Rose (1988) and Shao (1989a, 1989b).

Today, CD measurement SEMs already have field emission (FE) guns (Ohtaka et al., 1985; Shimada et al., 1986). The trade-off for this was the need for an improved vacuum. As the effective source size of the FE tip is already very small (≈20 nm), this means only a very slight electron optical reduction of the source size is needed in the column. There is a high sensitivity to vibrations and stray electromagnetic fields with FE guns. Preventive measures also have their price.

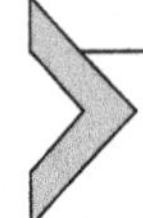

2. CALCULATING ELECTROSTATIC AND MAGNETIC FIELDS

Starting with a preliminary design for an electron optical component, the first step in the process of electron optical simulation is to calculate its field. There are only a few cases where this step can be dispensed with, where the electron optical characteristics are already available in terms of formulae, graphs or tables as a function of the geometric design parameters, voltages or currents of the component.

2.1 Analytical Methods

As personal computers (PCs) are widely available today and are capable of solving simple three-dimensional potential problems within an acceptable time using the techniques of numerical analysis, the reader may think we are wasting our time discussing analytical methods. This is an unwarranted assumption, however, because even today there are still good reasons for using analytical methods provided this can be done in a practical way. If an analytical solution, even an approximate one, is available, then we can see immediately how the potential or field distributions vary as a function of the system parameters. If numerical methods are used, this physical insight can only be obtained by altering the system parameters a number of times. It is also important to have some way of checking numerical field calculations in special cases against approximate analytical solutions that are already known (e.g. on the optical axis or asymptotic field decay). Another reason for considering analytical methods of calculating fields is the next step involved in the electron optical simulation, the calculation of electron trajectories. As the usual numerical methods of calculating fields (e.g. finite differences, finite elements) only provide values for the potential at discrete points, precise ray tracing, even using extra two- or three-dimensional interpolation of potential and field values, is often impossible. This type of error, which occurs when trajectories are calculated numerically, does not arise when the field can be expressed in an analytical form. This does not necessarily mean that it has to be calculated analytically. Three cases are of interest in this context.

The first case is when the spatial field of a magnetic lens with rotational symmetry is calculated using the finite element method (FEM) or the finite difference method (FDM). If the field values on the axis are now fitted using a function that gives the typical form of solution for a magnetic lens along its axis, we can obtain an analytical, i.e. continuous, solution for the field over the whole space by means of analytical continuation using series expansion or a closed-form solution of the integral representation. We will return to this case later. The second case is when the potential of an electrostatic lens has been calculated using the FEM or FDM and the paraxial trajectories of the first order and the spherical aberration of the third order are required. Due to the limited number of nodes in the FEM program or the limited number of mesh points in the FDM program, the axial values of the potential may be too widely spaced for further calculations. This is because the first and second derivatives of the potential along the axis are also required for further calculations. It would still be possible to resort to some form of numerical interpolation, but it is better to fit the axial values to the type of function which would be encountered in cases with an analytical solution. It is then possible to obtain a sufficient number of additional sample points and avoid the "coarsening" effect of numerical differentiation by using analytical differentiation.

The third case is a special numerical method of performing field calculations, the charge simulation method (CSM), which strictly speaking is not a method of numerical analysis at all, and is often classified by Kasper (e.g. Kasper, 1982) as an analytical method for calculating fields. In the case of a lens with rotational symmetry, each (thick) electrode is replaced by a sufficient number of circular apertures and charged rings whose potential in space can be calculated analytically and which are all located inside the electrodes (see Fig. 9). The positions and substitute charges are now chosen so that the superposed field of all the apertures and rings gives a constant potential profile at the actual surface of the electrode. This completes the numerical analysis itself, and the potential and field at any point in space can be obtained by summing the analytical expressions for all the circular apertures and rings in the course of ray tracing. In spite of the partially analytical aspects of the CSM, we still consider it to be a method of numerical analysis.

In the following sections on analytical methods for calculating fields, we will only collect a number of helpful formulae and examples. The following references provide more information on advanced analytical methods (Ollendorf, 1932, 1952; Weber, 1950; Smythe, 1950; Buchholz, 1957; Wendt, 1958; Moon & Spencer, 1961; Durand, 1964, 1966a, 1968; Küpfmüller, 1965; Simonyi, 1966; Jackson, 1975). The three books by Durand (1964, 1966a, 1968), in particular, contain many solved problems.

Planar Multipole Representation

Electron optical systems with straight axes are of prime importance for e-beam testing and inspection. For systems of this kind, the expansion of the scalar potential using planar multipoles or Fourier components, i.e. multiplicities about the straight axis (Rose, 1966/1967), has proved particularly effective:

$$\varphi = \sum_{\nu=0}^{\infty} \varphi_\nu \quad \text{for the electrostatic case} \tag{1a}$$

$$\psi = \sum_{\nu=0}^{\infty} \psi_\nu \quad \text{for the magnetostatic case} \tag{1b}$$

For each multipole component φ_ν or ψ_ν, which are, of course, functions of all three Cartesian coordinates x, y, z, there is just one (complex) axis function $\Phi_\nu(z)$ or $\Psi_\nu(z)$, where the z-axis, as is usual, lies along the optical axis. This means that, if the axial multipole strength is known, the space potential can be obtained by means of analytical continuation for each multipole component. The analytical continuation can be obtained in terms of a power series expansion or of integrals.

In the following discussion, we will often express the Cartesian coordinates x, y that are normal to the axis in terms of a complex variable,

$$w = x + \mathrm{i}y \quad \text{and} \quad \overline{w} = x - \mathrm{i}y \tag{2}$$

sometimes we will also use cylindrical coordinates

$$r = |w| = \sqrt{x^2 + y^2} \quad \text{and} \quad \vartheta = \frac{1}{2\mathrm{i}} \ln \frac{w}{\overline{w}} = \arctan \frac{y}{x} \tag{3}$$

For the electric field $\mathbf{E} = -\nabla\varphi$, the following relations in w, z coordinates can be derived:

$$E_w = E_x + \mathrm{i}E_y = -2\frac{\partial\varphi}{\partial\overline{w}} \qquad E_z = -\frac{\partial\varphi}{\partial z} \tag{4}$$

For the magnetic flux density $\mathbf{B} = -\nabla\psi$, the same equations but with ψ substituted for φ apply.

Power Series Expansion. If we expand the multipole components, each of which satisfies the Laplace equation, in a power series using the off-axis coordinates w and $\overline{w}$, we obtain the following expression in the electrostatic case (Rose, 1966/1967, 1968):

$$\begin{aligned}\varphi_\nu &= \sum_{\lambda=0}^{\infty} (-1)^\lambda \frac{\nu!}{\lambda!(\lambda+\nu)!} \left(\frac{w\overline{w}}{4}\right)^\lambda \mathrm{Re}\left\{\overline{w}^\nu \frac{\partial^{2\lambda}\Phi_\nu(z)}{\partial z^{2\lambda}}\right\} \\ &= \sum_{\lambda=0}^{\infty} (-1)^\lambda \frac{\nu!}{\lambda!(\lambda+\nu)!} \left(\frac{r}{2}\right)^{2\lambda} r^\nu \left\{\frac{\partial^{2\lambda}\Phi_{\nu c}(z)}{\partial z^{2\lambda}} \cos\nu\vartheta + \frac{\partial^{2\lambda}\Phi_{\nu s}(z)}{\partial z^{2\lambda}} \sin\nu\vartheta\right\}\end{aligned} \tag{5}$$

with $\Phi_0 = \overline{\Phi}_0 = \Phi_{0c} = \Phi$ for the component with rotational symmetry, where $\nu = 0$, $\Phi_\nu = \Phi_{\nu c} + i\Phi_{\nu s}$ for $\nu \geq 1$ and Re = real part.

For the multipole components ψ_ν of the magnetic potential ψ an expression of the same form, with $\Psi_\nu(z)$ substituted for $\Phi_\nu(z)$, applies. The component with rotational symmetry has the following expression added to it:

$$\Psi_0'(z) = \partial\Psi_0(z)/\partial z = \overline{\Psi}_0' = \Psi_{0c}' = -B_z(w = 0,\ z) = -B(z) \tag{6}$$

In the magnetic case, the flux density $\mathbf{B}$, given by the equation $\mathbf{B} = \nabla \times \mathbf{A}$ can also be derived from a vector potential $\mathbf{A}$. With the gauge $A_z = 0$, the expression

$$\frac{\partial A}{\partial z} = 2i\frac{\partial \psi}{\partial \overline{w}} \quad \text{with } A = A_x + iA_y = \sum_{\nu=0}^{\infty} A_\nu \tag{7}$$

can be derived. The following power series expansion is then obtained for the multipole components A_ν of the vector potential (Rose, 1968)

$$\begin{aligned} A_\nu = i\sum_{\lambda=0}^{\infty}(-1)^\lambda \frac{\nu!}{\lambda!(\lambda+\nu)!}\left(\frac{w\overline{w}}{4}\right)^\lambda \Bigg\{ &(\nu+\lambda)\overline{w}^{\nu-1}\frac{\partial^{2\lambda-1}\Psi_\nu(z)}{\partial z^{2\lambda-1}} \\ &+ \lambda w^\nu \overline{w}^{-1}\frac{\partial^{2\lambda-1}\overline{\Psi}_\nu(z)}{\partial z^{2\lambda-1}} \Bigg\} \end{aligned} \tag{8}$$

If very long electrodes are arranged along the optical axis (i.e. planar case with $\partial/\partial z = 0$) an expansion using planar multipoles (5) is equivalent to an expansion using harmonic polynomials, which is well-known from the theory of analytic functions (Durand, 1966a; Szilagyi, 1988). The following is a list of the polynomials up to and including the third degree:

$$\left.\begin{array}{ll} 1 & 0 \\ x = r\cos\vartheta & y = r\sin\vartheta \\ x^2 - y^2 = r^2\cos 2\vartheta & 2xy = r^2\sin 2\vartheta \\ x^3 - 3xy^2 = r^3\cos 3\vartheta & 3x^2y - y^3 = r^3\sin 3\vartheta \end{array}\right\} \tag{9}$$

Fig. 1 shows arrangements of electrodes that produce a pure dipole field, quadrupole field or hexapole field when $\partial/\partial z = 0$. The advantage of an expansion using planar multipoles, or harmonic polynomials in the special case, is that *each* of these functions satisfies Laplace's equation individually. This is not the case when Cartesian polynomials are used for the expansion (Glaser, 1952) as even each of the polynomials $x^2, y^2, x^3, x^2y, xy^2$, or y^3 in

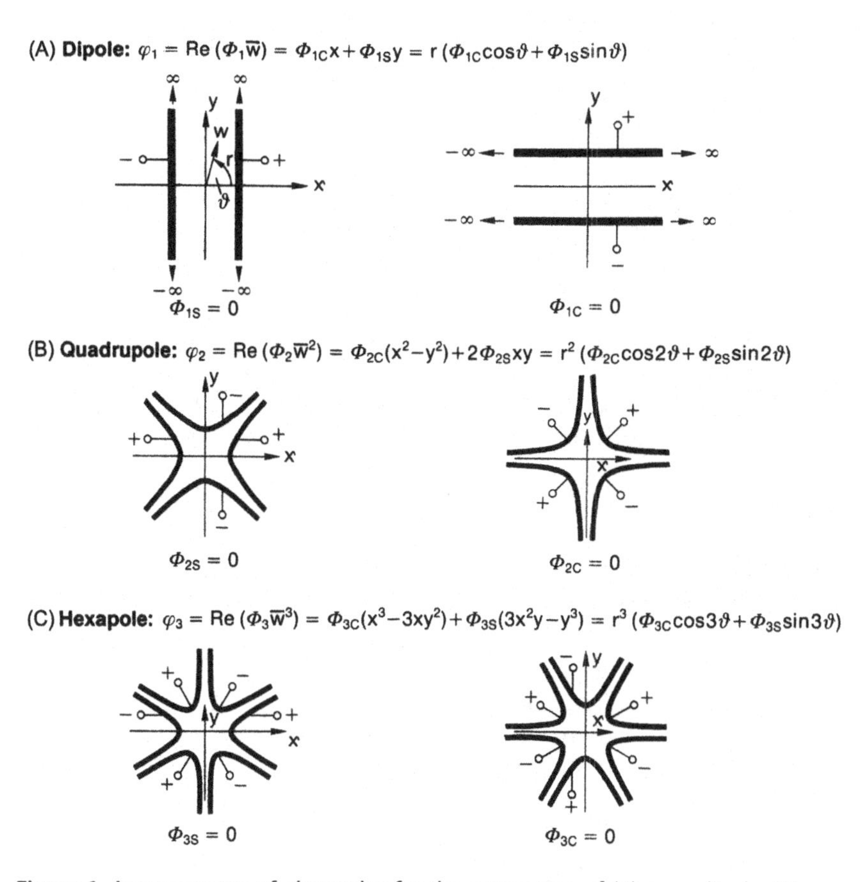

Figure 1 Arrangements of electrodes for the generation of (A) pure dipole, (B) pure quadrupole, and (C) pure hexapole fields. *From Plies (1982a).*

the two-dimensional case does not satisfy Laplace's equation individually. In the general case, complicated relationships between the coefficients of the expansion which depend on z are obtained. Moreover, the expansion using planar multipoles is more suitable from the point of view of aberration theory.

The following comments on expansions using planar multipoles have been translated directly from the German language paper of Heijnemans, Nieuwendijk, and Vink (1980/1981):

In recent years, the authors have made use of the multipole expansion for magnetic deflection fields in the course of development work on deflection coils. This has made insights into the relationships between coil/field and field/deflection considerably clearer, thus making the development of deflection

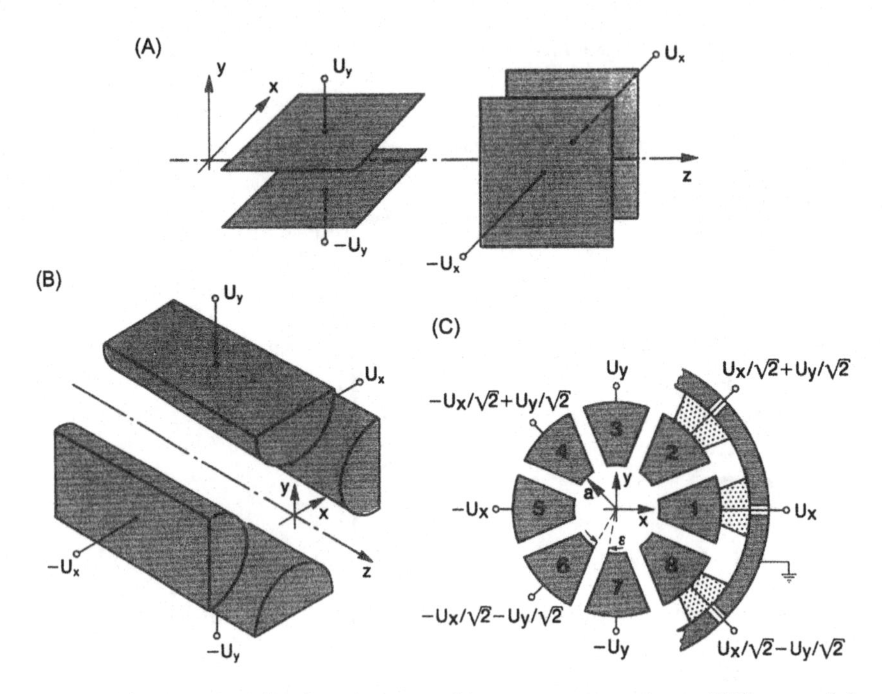

Figure 2 Electrostatic deflection elements with two symmetry planes. (A) Two parallel-plate capacitors for spatially separated *x*- and *y*-deflection; (B) four-pole, and (C) eight-pole deflection system for superimposed *x*- and *y*-deflection. *From Plies (1982a).*

units more like analyzing a difficult but fascinating chess game rather than toiling through a mass of unenlightening computations that had to be tackled previously.

If the potential of the deflection elements has two planes of symmetry (symmetrical and antisymmetrical) as shown in Fig. 2, all the even multipole components of the deflection field vanish, i.e. $\varphi_{2\nu} = \Phi_{2\nu} = 0$. In the case of the eight-pole element shown in Fig. 2C, the additional condition $\Phi_3(z) = \Phi_5(z) = 0$ also applies (Ohiwa, 1985).

Integral Representation. The power series expansion (5) is particularly useful when the complex multipole strengths $\Phi_\nu(z)$ are known and the potential of the multipole components $\varphi_\nu(w, \overline{w}, z)$ in the vicinity of the axis is required. If there is also interest in further off-axial regions or even in the geometry of the electrodes associated with $\Phi_\nu(z)$ (or the pole pieces or the windings of iron-free coils in the magnetic case) the associated integral representation may be more convenient than the power series expansion. The formula given by Rose (1966/1967, 1987) is

$$\varphi_\nu = \frac{1}{2\pi}\frac{4^\nu(\nu!)^2}{(2\nu)!}\mathrm{Re}\left\{\overline{w}^\nu \int_0^{2\pi} \Phi_\nu(z + ir\sin\alpha)\cos^{2\nu}\alpha\, d\alpha\right\}$$

$$= \frac{1}{2\pi}\frac{4^\nu(\nu!)^2}{(2\nu)!} r^\nu \int_0^{2\pi} \{\Phi_{\nu c}(z + ir\sin\alpha)\cos\nu\vartheta$$

$$+ \Phi_{\nu s}(z + ir\sin\alpha)\sin\nu\vartheta\}\cos^{2\nu}\alpha\, d\alpha \qquad (10)$$

The formula when there is symmetry of rotation ($\nu = 0$) was first derived by Laplace (see the comment in Scherzer, 1980). Eq. (10) can be transformed into Eq. (5) (see Rose, 1966/1967, 1987).

Solutions to Eq. (10) are well known in the literature if rotational symmetry is present ($\nu = 0$), e.g. for the following axis functions:

(α) $\Phi = Az^2$ (Szilagyi, 1988)

(β) $\Phi = \dfrac{az + b}{1 + z^2}$ (Glaser, 1956)

(γ) $\Phi = Ae^{-z/a}$ (Glaser, 1952)

(δ) $\Phi = Aa + B(a^3 + \frac{3}{2}az^2) + [Az + \frac{3}{2}(z^3 + a^2 z)]\arctan\left(\frac{z}{a}\right)$ (Scherzer, 1980)

(ε) $\Psi_0 = -B_0 d \arctan(z/d)$ or $B = \dfrac{B_0}{1 + (z/d)^2}$ (Glaser, 1956; Rose, 1987)

For $\nu = 2$, Rose (1987) solved Eq. (10) for the magnetic quadrupole field strength given by

(ζ) $\Psi_2 = \Psi_{2c} = \dfrac{A}{[1 + (z/a)^2]^2}$

Case (α) describes the axial potential in the vicinity of a saddle point on the axis of rotation. Case (β) is the bell-shaped axial potential distribution (up to a constant) that is characteristic of a three-aperture lens (immersion lens $a \neq 0$, Einzel lens $a = 0$). The axial potential function given for case (δ) is associated with a transparent foil lens with a circular aperture as suggested by Scherzer (1980) to correct third-order spherical aberration. Scherzer continued this axial potential function in space and calculated the necessary influence charges on the foil. Case (ε) is known as Glaser's bell-shaped field, which has been of great importance in the electron optics of magnetic lenses as it allows an analytical solution of the paraxial ray equation (Glaser, 1952) to be found. A similar $B(z)$ curve is given by a current loop, which means that (ε) does not model the axial field of unsaturated, magnetic pole piece lenses very well. Ruska (1965a, 1965b) has found that the focal properties of saturated pole piece lenses can be described using Glaser's bell-shaped field. Superficially, therefore, it would seem that the bell-shaped field would be of little significance for low-voltage applications because, as a rule, the pole pieces are unsaturated. This, however, does not take unconventional pole piece designs into account, e.g. the field set up by a single pole piece lens may have the same form as Glaser's bell-shaped field (Juma, 1986). It has also been shown by Tretner (1959) and Moses (1973) that Glaser's bell-shaped field is also a good approximation to the field distribution with minimal spherical aberration of the third order. It is therefore still of interest, even for low-voltage applications. Plies and Schweizer (1987) have used the analytical continuation of Glaser's asymmetric bell-shaped field (Glaser, 1952) to find the off-axis secondary electron trajectories in a compound spectrometer objective lens used for e-beam testing. To avoid problems with the unrealistically slow decrease of Glaser's field, they cut it off below a certain axial field value of B.

Using the quadrupole strength of case (ζ), Rose (1987) found a three-dimensional scalar magnetic potential distribution for a perfect magnetic quadrupole with an exact treatment of the fringing fields. This quadrupole field can be generated by four flat coils.

Fourier–Bessel Series

In potential theory, the method of separation of variables is often used to solve Laplace's equation in cylindrical coordinates (r, ϑ, z), and the solutions can be expressed in terms of the Fourier–Bessel series below:

$$\left.\begin{aligned}&\varphi = \sum_{\nu=0}^{\infty} \varphi_\nu\\&\text{with}\\&\varphi_\nu = \sum_{\lambda=0}^{\infty} I_\nu(k_\lambda r)\left[A_\lambda \cos(k_\lambda z) + B_\lambda \sin(k_\lambda z)\right]\left[C_\nu \cos(\nu\vartheta) + D_\nu \sin(\nu\vartheta)\right]\end{aligned}\right\} \tag{11}$$

where I_ν is the modified Bessel function of the first kind and νth order. The constants A_λ, B_λ, k_λ, C_ν, and D_ν must be chosen to satisfy the boundary

(A) **Plane z = constant:**

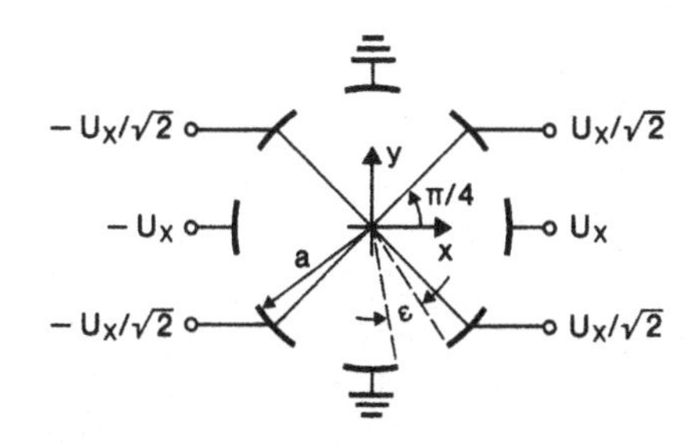

(B) **Section y = 0:**

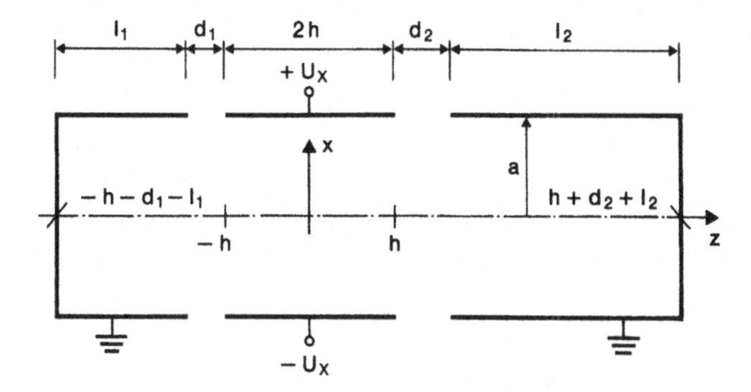

Figure 3 Geometry of an electrostatic eight-pole deflector between two closed cylindrical tubes, and voltage supply for deflection in the x-direction. *From Plies and Elstner (1989b).*

conditions of the potential problem. Eq. (11) is not the most general solution to Laplace's equation (see Durand, 1966a) but it is quite sufficient for electron optical systems with straight axes. In Eq. (11), it is sometimes convenient to replace summation over λ by integration. In passing, it should be mentioned that Eq. (11) can be expressed in terms of the multipole components φ_ν in Eq. (5) or Eq. (10) when the modified Bessel function I_ν in Eq. (11) is replaced by its Taylor series expansion or its integral representation, respectively.

The Fourier–Bessel series (11) was used to calculate two-tube and multitubular lenses by Bertram (1940), Frobin (1968), Glikman, Kel'man, and Nurmanov (1974), and van der Merwe (1978, 1979). Franzen (1984) applied it to quadrupole-type Klemperer lenses (Klemperer, 1953) in cathode ray tubes. Plies and Elstner (1989b) used this method to calculate the axial deflection field distribution (=dipole strength) of the eight-pole deflector shown in Fig. 3. It was assumed that the potential in both the azimuthal gaps

(Fig. 3A) and in the axial gaps (Fig. 3B) varies linearly. Oku and Fukushima (1986) have used the Fourier–Bessel series to calculate the spatial potential in Schlesinger's deflectron (Schlesinger, 1952, 1956), which is a special electrostatic pattern yoke.

Lens Fields

In this section, we intend to present a few analytical field models for the axial distribution $\Phi(z)$ or $B(z)$ of lens fields with rotational symmetry. One kind of field model is the result of analytical field calculations which make certain approximating assumptions, e.g. for the gap/bore ratio of an iron-shielded magnetic lens. We could refer to these field models as Laplacian because they satisfy Laplace's equation. The second kind of field model only provides the functional form which can be used to give good approximations to real lens fields by choosing suitable values for the free parameters of the field model. Basically, Laplacian field models are to be preferred, because when the axial curve is subjected to analytical continuation in space (by power series or integral representation), no problems (singularities, exotic electrodes) usually arise as is often the case with the second kind, the formal field models. The advantage of a few model fields of the second type is that they can be used to derive analytical solutions for paraxial trajectories, or even for the image aberration integrals. In this sense, Glaser's model field of the magnetic lens has made considerable contributions to our understanding of the characteristics of this lens, although, in many cases, it does not model the axial field distribution very well (Lenz, 1982). Let us now return to our discussion of Laplacian model fields. As a rule, they contain the geometrical lens parameters and the lens voltages or the magnetic lens excitation. If these real parameters are replaced by formal fitting parameters, these field models can give very good approximations for real lens fields that do not even have to satisfy the original assumptions underlying the approximations made. We will return to these matters later.

Electrostatic Lenses. To model electrostatic lenses, charged coaxial rings or apertures can be used in the same way as is done for the CSM. This is because the potential distribution in space for these rings or apertures can be given in closed form. As is well known, in the case of the charged ring this involves the use of the complete elliptic integral of the first kind (Kasper, 1982), but the axial distribution can be expressed simply as

$$\Phi(z) = \frac{Q}{4\pi\varepsilon_0\sqrt{a^2 + z^2}}$$

where Q is the charge on the coaxial ring of radius a which is situated in the plane defined by $z=0$ and ε_0 is the electric constant.

The space potential of a thin circular coaxial aperture has been derived using oblate spheroidal coordinates, e.g. by Brüche and Scherzer (1934), Durand (1966a), El-Kareh and El-Kareh (1970a), and Kasper (1982). Here we will only state the axial distribution $\Phi(z)$ and the potential distribution $\varphi(r, 0)$ in the coaxial aperture lying in the plane defined by $z=0$:

$$\Phi(z)=\varphi(0,z)=\varphi_A-\frac{1}{2}(E_1+E_2)z+\frac{a}{\pi}(E_1-E_2)\left[1+\frac{z}{a}\arctan\left(\frac{z}{a}\right)\right] \tag{12a}$$

$$\varphi(r,0)=\varphi_A+\frac{1}{\pi}(E_1-E_2)\sqrt{a^2-r^2} \quad \text{for } r\leqslant a \tag{12b}$$

φ_A is the potential and a the radius of the aperture electrode which is positioned between two uniform fields, $E_1=-\Phi'(-\infty)$ and $E_2=-\Phi'(+\infty)$.

Three functions of the same type as Eq. (12a) can be used to describe the three-aperture lens (Regenstreif, 1951; Glaser, 1952). This model can also be used to calculate approximations to the axial potential distribution of an Einzel lens with a thick center electrode, if four (El-Kareh & El-Kareh, 1970a) or even five substitute apertures are used (Kanaya, Kawakatsu, Yamazaki, & Sibata, 1966). However, if the thickness of the center electrode is greater than its bore diameter, the five-electrode approximation is not sufficiently accurate (Yamazaki, 1979). For Einzel lenses with thicker center electrodes and thickness/bore diameter $\geqslant 1.5$, Yamazaki gives a relatively simple axial distribution $\Phi(z)$, which also contains arctan functions.

In applications for e-beam testing, grids of the retarding-field spectrometer are often inserted in the beam path of SEs. The meshes of these grids are usually square, but the central axial mesh through which the primary beam passes is often circular. This reduces the axial astigmatism of the PE probe. There is an electric field on at least one side of a solid mesh, so that in the central mesh the primary beam is subject to a lens effect. Eq. (12a) can also be used (Plies & Elstner, 1989b) to take this effect approximately into account.

Glaser and Schiske (1954a) have used the potential function

$$\Phi=U+\frac{U_L}{1+(z/d)^2} \tag{13}$$

to describe a typical electrostatic Einzel lens (three-electrode lens with the plane of symmetry given by $z=0$). From Eq. (13) we can see that $U=$

$\Phi(-\infty) = \Phi(+\infty)$ is the potential at very large distances from the lens and is equal to the potential of the external electrodes. U_L is the *effective* lens voltage because $U + U_L$ is the potential at the saddle point and not at the center electrode. Finally, d is the half-width of the second term in Eq. (13). The advantage of this type of potential distribution model is that it allows a closed-form solution for the paraxial trajectories, the cardinal elements and the image aberration integrals as a function of the lens strength $\varkappa^2 = -U_L/U$ (Glaser & Schiske, 1954a, 1954b, 1955; Glaser, 1956). However, Eq. (13) gives an asymptotic approximation to U which is too slow when U, U_L, and d are chosen to fit a real potential distribution (e.g. see Glaser, 1956, Fig. 33).

Kanaya and Baba (1978) used the model potential

$$\Phi(z) = \Phi(0)\exp\left[K_0 \arctan(z/a)^m\right] \tag{14}$$

for the symmetric Einzel lens (plane of symmetry defined by $z = 0$). With this four-parameter model potential, we can obtain more realistic asymptotic behavior than by using Glaser and Schiske's model potential given by Eq. (13). Over and above this, the fundamental trajectories in the form of hypergeometric functions can be given for the model field described by Eq. (14). Analytic expressions for the focal properties, the spherical and chromatic aberrations can also be given. The work of Kanaya and Baba (1978) also contains many diagrams for designing practical three-electrode lenses.

We have hitherto examined model fields only for Einzel lenses (unipotential lenses), but shall now turn to immersion lenses (bipotential lenses). With this type of lens, the potentials on the right and left of the lens field itself are different, $U_1 = \Phi(-\infty) \neq U_2 = \Phi(+\infty)$. A simple example illustrating this type of lens is the geometrically symmetric, two-tube lens where both tube electrodes have the same diameter D with a gap of width S between them. Assuming a linear potential distribution in the gap, the Dirichlet potential problem can be solved using the previously mentioned methods of separation of variables when cylindrical coordinates are used. This solution can then be expressed in terms of the Fourier–Bessel integral (e.g. see El-Kareh & El-Kareh, 1970a; Grivet, 1972). It is then found that the axial potential distribution is very well described by the function

$$\left.\begin{aligned}\Phi &= \frac{U_1+U_2}{2} + \frac{U_2-U_1}{2}\frac{D}{\omega S}\ln\frac{\cosh[\omega/D(z+S/2)]}{\cosh[\omega/D(z-S/2)]}\\ \text{with}&\\ \omega &= \frac{4}{\pi}\int_0^{\infty} \mathrm{d}k/I_0(k) = 2.6367\end{aligned}\right\} \quad (15)$$

This is referred to as Bertram's model (Bertram, 1940, 1942) and is very accurate in spite of the assumption of a linear decrease in gap potential. Even when $S = D$, the results are sufficiently accurate for many applications. We must bear in mind that an error of 20% for spherical aberration, due to the field model, can be accepted for preliminary calculations because as far as high-resolution (and medium-voltage) electron microscopy is concerned, for example, the resolution only varies as the fourth root of the spherical aberration. For $S \ll D$, Eq. (15) takes on the form of the Gray model (Gray, 1939):

$$\Phi = \frac{U_1+U_2}{2} + \frac{U_2-U_1}{2}\tanh\left(\frac{\omega z}{D}\right) \quad (16)$$

The Fourier–Bessel series can be used successfully even for multitubular lenses with a common radius. In this case, it should be assumed that the potential distribution varies at least as fast as a linear function and does so in all gaps (Frobin, 1968). This is because the Fourier coefficients then vary as $1/k_\lambda^2$, where k_λ is the index. Apart from the better approximation from the physical point of view, the series converges better than it would if potential jumps were assumed (coefficients vary as $1/k_\lambda$).

Kanaya and Baba (1977) have stated a three-parameter axial potential distribution for immersion lenses with various electrode configurations. In this case, the paraxial trajectories can be expressed in terms of hypergeometric functions. Even the focal length and the spherical and chromatic aberrations were formulated analytically and illustrated by a large number of useful diagrams. This axial model potential subsumes two simpler two-parameter model potentials that were previously derived by Hutter (1945) and Kanaya, Kawakatsu, and Miya (1972).

For low-voltage applications, electrostatic lenses, in particular immersion lenses, have recently become very fashionable. This trend will probably become even more marked and this is why the discussion of axial potential models for electrostatic lenses was relatively wide-ranging.

Magnetic Lenses. Lenz (1982) has already made a compilation of analytical field models for magnetic lenses and discussed them exhaustively. We will

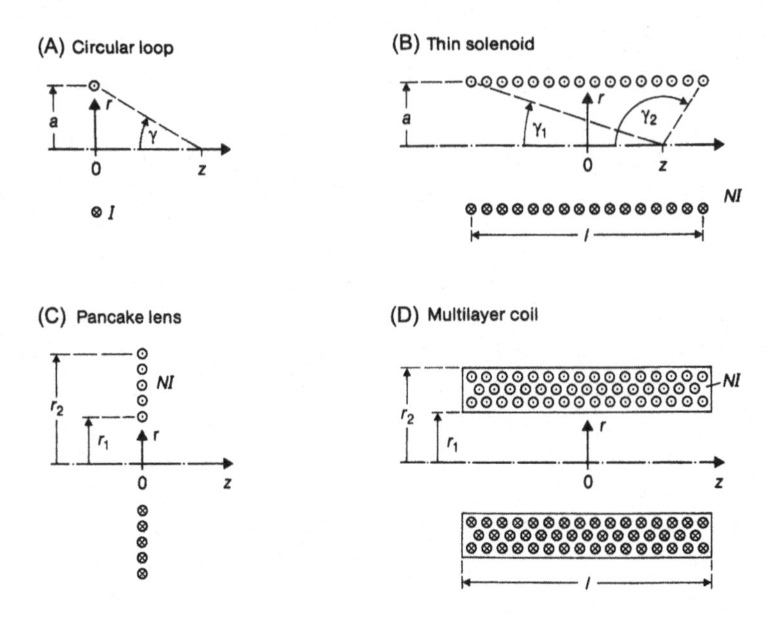

Figure 4 Four arrangements of circular currents generating a rotationally symmetric magnetic field.

not describe the models treated by Lenz in any great detail, but will go on to mention a few additional models. The axial field distribution $B(z)$ for most magnetic lenses is bell shaped, like the field distribution $\Phi' = \partial\Phi/\partial z$ for the electrostatic immersion lenses. Some magnetic models are therefore analogous to electrostatic models that have been discussed previously. Before we deal with symmetric and asymmetric iron-shielded magnetic lenses, a few formulae for the axial distribution $B(z)$ in cylindrical coils will first be stated. Fig. 4 shows some typical cylindrical coils and the relevant symbols used in their description. As usual, the symbol μ_0 denotes the magnetic constant.

Simple circular current (Fig. 4A):

$$B = \frac{\mu_0 I}{2}\frac{a^2}{(a^2 + z^2)^{3/2}} = \frac{\mu_0 I}{2a}\sin^3\gamma \tag{17}$$

Monolayer solenoid (Fig. 4B)

$$B = \frac{\mu_0 NI}{2l}(\cos\gamma_1 - \cos\gamma_2) \tag{18}$$

with

$$\cos\gamma_1 = (l/2+z)\left[(l/2+z)^2+a^2\right]^{-1/2}$$
$$\cos\gamma_2 = -(l/2-z)\left[(l/2-z)^2+a^2\right]^{-1/2} \tag{19}$$

Pancake lens (Fig. 4C):

$$B=\frac{\mu_0 NI}{2(r_2-r_1)}\left\{\frac{r_1}{(r_1^2+z^2)^{1/2}}-\frac{r_2}{(r_2^2+z^2)^{1/2}}+\ln\frac{r_2+(r_2^2+z^2)^{1/2}}{r_1+(r_1^2+z^2)^{1/2}}\right\} \tag{20}$$

Multilayer coil (Fig. 4D):

$$B=\frac{\mu_0 NI}{2(r_2-r_1)l}\left\{\left(z+\frac{l}{2}\right)\ln\frac{r_2+[r_2^2+(z+l/2)^2]^{1/2}}{r_1+[r_1^2+(z+l/2)^2]^{1/2}}\right.$$
$$\left.-\left(z-\frac{l}{2}\right)\ln\frac{r_2+[r_2^2+(z-l/2)^2]^{1/2}}{r_1+[r_1^2+(z-l/2)^2]^{1/2}}\right\} \tag{21}$$

The derivation of these formulae can be found in many textbooks dealing with potential theory or electron optics, e.g. Szilagyi (1988). For more on the iron-free pancake lens, see Mulvey (1982).

Let us now turn our attention to magnetic pole piece lenses where the coil which has a lens excitation of NI is iron shielded, so that the magnetomotive force when the iron is unsaturated drops in the axial pole piece gap. If the lens is symmetrical to the plane $z=0$ and the pole piece gap S is small compared with the bore diameter D, the axial flux density can be described by the Gray Model (Gray, 1939),

$$B=\frac{\mu_0\omega NI}{2D}\frac{1}{\cosh^2(\omega z/D)} \tag{22}$$

or the Grivet–Lenz model field (Grivet, 1951, 1952; Lenz, 1951),

$$B=\mu_0\frac{2\alpha_1}{\pi}\frac{NI}{D}\frac{1}{\cosh(2\alpha_1 z/D)} \tag{23}$$

The value of ω in Eq. (22) is 2.6367, as in Eq. (15). Eq. (22) is the z-derivative of the electrostatic model represented by Eq. (16). The value of α_1 in Eq. (23) is 2.4048, the first zero of the zeroth-order Bessel function of the first kind $J_0(z)$. For both field models, the decrease in B at large $|z|$ follows almost the same exponential function, but an analytical solution for the fundamental trajectories can be derived from the Grivet–Lenz model.

If it is no longer true that $S \ll D$, it is better to use the Bertram (1940, 1942) model,

$$B = \mu_0 \frac{NI}{2S} \frac{\sinh(\omega S/D)}{\cosh[(\omega/D)(z+S/2)]\cosh[(\omega/D)(z-S/2)]} \tag{24}$$

As $S \to 0$ in Eq. (24), Eq. (22) is obtained. Eq. (24) is the derivative with respect to z of the electrostatic model (15). When the Gray model (22) was derived with $S \approx 0$, a jump in the magnetic potential in the pole piece gap was assumed, while a linear decrease was assumed for the Bertram model. Grivet (1972) compared the values given by the Bertram model for $0 < S/D < 2$ with the measurements performed by van Ments and Le Poole (1947) and found excellent agreement. Grivet also found very good agreement with numerical calculations performed by Liebmann (1951), Liebmann and Grad (1951), and Lenz (1950a).

Lenz (1982) describes a convolution technique to go from field models for $S \ll D$ to improved field models for any value of S/D. In this sense, the Bertram model (24) is the convoluted Gray model (22). (Eq. (24) can be derived from the three formulae given by Lenz, 1982.) The convolution of the Grivet–Lenz field gives

$$B = \frac{2\mu_0 NI}{\pi S} \arctan \frac{\sinh(\alpha_1 S/D)}{\cosh(2\alpha_1 z/D)} \tag{25}$$

When the field of a current loop given by Eq. (17) undergoes convolution, the field of a monolayer solenoid given by Eq. (18) is obtained. According to Durandeau (1957), Eq. (18) with $l = S$ and $a = D/3$ is a good approximation for an unsaturated pole piece lens, provided that $0.5 < S/D < 2$. This approximation is also referred to as the equivalent solenoid lens by Mulvey and Wallington (1973). The advantage of Durandeau's model is that the electron optical properties are only a function of the length

$$L = \sqrt{S^2 + (2D/3)^2} \approx \sqrt{S^2 + 0.45D^2} \tag{26}$$

where L is equal to the diagonal of the equivalent solenoidal lens.

The Glaser model field (Glaser, 1941, 1952) is suitable for describing symmetrical magnetic lenses near saturation

$$B = \frac{B_0}{1 + (z/d)^2} \tag{27}$$

B_0 is the maximum flux density and d the half-width. The greatest advantage of this field distribution is that it not only gives an analytical solution for the paraxial trajectory equation but also for the image aberration integrals of the first and third order. Moreover, in this case Newton's equation is satisfied for all object and image positions. In order to "cash in" on all these advantages, the Glaser bell-shaped field must be adapted to a field distribution that has been calculated (or measured) in a different way. If, as Lenz (1982) did, one adapts the half-width and the lens excitation,

$$NI = \frac{1}{\mu_0} \int_{-\infty}^{+\infty} B(z)\mathrm{d}z \tag{28}$$

Glaser's bell-shaped field comes off badly in comparison because of its slow asymptotic decrease. Adapting Glaser's model by Grivet's procedure (Grivet, 1972) using

$$\int_{-\infty}^{+\infty} \left(B^2 - B^2_{\mathrm{Glaser}}\right)\mathrm{d}z = 0 \tag{29}$$

shows Glaser's bell-shaped model in a better light. It is also possible to adapt maximum induction and half-width. To prevent the excessive excitation that would now occur, one can, for example, arrange that the field is zero from a certain z-value $\pm z_{\mathrm{L}}$ in using ray tracing based on the Glaser bell-shaped field with analytical continuation, e.g. by using

$$\int_{-z_{\mathrm{L}}}^{+z_{\mathrm{L}}} B_{\mathrm{Glaser}}\,\mathrm{d}z = \int_{-\infty}^{+\infty} B\,\mathrm{d}z = \mu_0 NI \tag{30}$$

Sometimes the generalized Glaser bell-shaped function

$$B = \frac{B_0}{[1 + (z/a)^2]^{\nu}} \tag{31}$$

with the half-width $d = a\sqrt{2^{1/\nu} - 1}$ is considered (Glaser, 1941; Lenz, 1950a; Glaser & Lenz, 1951). With $\nu > 1$, a more rapid decrease in the field can be modeled, but an analytic solution for the fundamental trajectories is only known for $\nu = \frac{3}{2}$ (see below). If a dimensionless shape factor is defined by $\mu_0 NI/(B_0 d)$, it is equal to π for $\nu = 1$, and to 2.61 for $\nu = \frac{3}{2}$, which corresponds to the field produced by a current loop. It is equal to 2.13 as ν tends to infinity. This shape factor is 2.27 for the Gray model and 2.39 for the Grivet–Lenz model (23). The shape factor for saturated iron is always greater than that for unsaturated iron, provided the lens geometry is the same.

Kanaya, Baba, and Ono (1976) have presented a three-parameter model that allows the expression of the fundamental trajectories in terms of hypergeometric functions. As it also satisfies the Newton equation, exact expressions for the cardinal points and the spherical and chromatic aberration could be derived. This model field subsumes the constant field, the Grivet–Lenz field, Glaser's bell-shaped field and the one and a half power field (current loop) as special cases.

It should also be mentioned that Typke (1972) reformulated the Bertram field (24) (using α_1 instead of ω) and introduced three new formal parameters, the maximum field B_0, the half-width d and a special shape parameter τ instead of $\mu_0 NI$, S, and D. He was able to show that this expression is associated with a Heun differential equation for the electron trajectories (Heun, 1889), and this form of the field subsumes the Glaser bell-shaped field, the Grivet–Lenz field, the Gray field and the "top-hat field" as special cases. Typke has also derived focal length and aperture aberrations of the third and fifth order for various shape parameters as an approximation for the short lens.

In many cases, the magnetic lens is not symmetric and there are two different bore diameters ($D_1 \neq D_2$) in the pole piece system. This, for example, applies to the pinhole lens which is often used as an objective lens in SEMs. Glaser's asymmetrical bell-shaped field (Glaser, 1941, 1952) can be used to describe asymmetric magnetic lenses near saturation:

$$\left.\begin{aligned} B &= B_0/\left[1+(z/d_1)^2\right] \quad \text{for } z<0 \\ B &= B_0/\left[1+(z/d_2)^2\right] \quad \text{for } z>0 \end{aligned}\right\} \tag{32}$$

Lenz (1982) has provided a critical discussion of this model field with associated warnings because the second and higher derivatives of $B(z)$ are discontinuous at $z=0$, i.e. Eq. (32) does not satisfy Laplace's equation at $z=0$. Nevertheless, Plies and Schweizer (1987) have used Eq. (32) and have provided an analytic continuation in space by means of the integral representation (10). Under these circumstances, derivatives do not occur formally in the expression, but the plane defined by $z=0$ remains a plane of discontinuity, i.e. at this point the magnetic flux jumps by $\approx \pi B_0(1/d_1^2 - 1/d_2^2)r^2/4$. When numerical ray tracing is being performed, it is in certain cases possible to integrate over this discontinuity. Then, provided that the radial coordinate r is still significantly smaller than the two half-widths d_1, d_2, and provided that these do not differ greatly, the error is always less than that due to the slow decrease encountered with the bell-shaped field. However, the latter can be reduced by setting $B(z)$ equal to zero for large $|z|$ (Plies & Schweizer, 1987).

There are other analytic models for asymmetrical lenses, e.g. by Durandeau (1957) and two by Lenz (1950b, 1982). Particularly worth mentioning is the simpler of the Lenz models:

$$B = \mu_0 \frac{NI}{2S}\left\{\tanh\left[\frac{\alpha_1}{2D_1}(2z+S)\right] - \tanh\left[\frac{\alpha_1}{2D_2}(2z-S)\right]\right\} \tag{33a}$$

As it expresses only the difference of two tanh functions and has a simple constant factor (when $D_1 = D_2$ and $\alpha_1 \to \omega$, Eq. (33a) becomes Eq. (24)). Using this function, unsaturated magnetic lenses with a simple asymmetrical pole piece geometry can be described. Moreover, all the higher derivatives of $B(z)$ can be expressed as polynomials in tanh functions (Plies, 1990b). Therefore, Eq. (33a) together with Eqs. (5) and (8) allow analytical continuation to be performed into space by a simple summation of series. In the case of an asymmetric lens with complicated pole pieces (i.e. D_1, D_2, and S cannot be defined directly) or at the onset of iron saturation, Eq. (33a) can be written in the form

$$B = P_1\left\{\tanh\left[P_2(2z+P_4)\right] - \tanh\left[P_3(2z-P_4)\right]\right\} \tag{33b}$$

where P_1 to P_4 are parameters used to fit the function $B(z)$ by the method of least squares to curves calculated by numerical analysis or determined by measurement. It is then possible to continue $B(z)$ in the form of Eq. (33b), as mentioned above, by the method of series expansion. High-precision ray tracing can then be carried out using this magnetic field which is continuous over space. In any case, this technique is more accurate than using discrete field values obtained by the FDM.

Double-gap pole piece lenses should also be mentioned at this juncture. These lenses are operated with excitations of opposite sense and can be used to compensate for image rotation. Dinnis (Dinnis, 1987, 1988; Dinnis & Khursheed, 1989) has suggested use of this kind of lens for the magnetic extraction of secondary electrons for e-beam testing. Baba and Kanaya (1979) have derived a model for this lens type which permits derivation of an analytical solution for the paraxial trajectory equation in the form of hyper-geometric functions. This publication contains a variety of diagrams relating to focal properties and aberrations, which can be used directly to design lenses of this kind. This three-parameter model field by Baba and Kanaya subsumes a special two-parameter model field of the form $B(z) = Cz/(1 + z^4/a^4)$ of Lenz (1956a).

In principle, it is more difficult to design a magnetic lens than an electrostatic one, as the former requires the determination not only of the internal geometry (parallel part of the lens gap and the bore diameters) but also of the whole magnetic circuit (pole piece region, lens core and external casing). Over and above this, the lens windings have also to be designed. The publications of Mulvey and Wallington (1973) and Riecke (1982) deal very well with the practical aspects of lens design. They also contain the semi-analytical method of Hildebrandt and Riecke (1966) for calculating the magnetic flux in the pole piece system. The Hildebrandt–Schiske procedure (Hildebrandt, 1954) for calculating the magnetic flux density in the lens core is also described.

Deflection Fields

Every SEM or similar device that uses the scanning principle is equipped with deflection elements to scan the electron probe over the specimen. In the fields of e-beam testing and e-beam inspection, as far as the author is aware, only magnetic beam deflection systems (single or multistage versions) are being used at present. This contrasts with e-beam lithography where electrostatic beam deflectors have been in use for a long time because of the higher deflection speed required in this application. In the near future, other contingencies may make it sensible to introduce electrostatic deflectors for e-beam testing and e-beam inspection. In these fields, detection of the SEs is increasingly being performed via the objective lens. This means that in the case of in-lens deflection for the PEs, the SEs must also pass through the magnetic deflection element, and so are in some way scanned with respect to the detector. If this effect causes problems and a two-stage pre-lens deflection cannot be implemented for other reasons, then combined electric/magnetic scanning may be used for the PE probe, in such a way that it acts like a Wien filter for the SEs, i.e. does not deflect them. We will consequently touch on electrostatic deflectors and ways of calculating their fields.

Electrostatic Deflectors. We will not discuss conventional electrostatic deflectors with parallel plates, as shown in Fig. 2A, as their principal deflection planes for x- and y-deflection do not coincide. The use of electrostatic deflection yokes with superposed fields for providing x- and y-deflection has intensified over the last 20 years. They consequently have a common center of deflection for both deflection directions. The pattern yokes and multipole yokes belong to this class of deflector. For more information on pattern yokes, whose main representative is the deflectron with zig-zag electrodes (Schlesinger, 1952, 1956) used in vidicons, reference should be made to the review articles by Hutter (1974) and Ritz (1979) and the newer work by Oku and Fukushima (1986), which contains a semi-analytical calculation of the spatial deflection field.

Multipole yokes or multipole deflectors employing longitudinal electrodes arranged on a cylindrical surface are suitable for electric beam scanning for SEMs; Fig. 2C shows an eight-pole deflector. The frequently used term "octupole deflector" has been avoided, as it could give rise to misleading associations with the multipole character of the deflection field. Because of the two-fold symmetry of the deflection field, all even terms in the expansion using planar multipoles, Eq. (5), vanish–including the octupole component.

When calculations of multipole deflector fields are described in the literature, it is often assumed that their length is effectively infinite (e.g. Kelly, 1977 or Wollnik, 1987). This is reasonable when preliminary, exploratory calculations are being made, but the second step, which considers the finite length of the arrangement and fringing fields, has to be taken at some time. In some cases, an analytical approach is taken to the second step. However, before going into this, we will discuss very long multipole deflectors with $\partial/\partial z \approx 0$. The potential of the homogeneous transverse field, as is generally known, is given by $\varphi = \Phi_{1c}x + \Phi_{1s}y = \Phi_{1c}r\cos\vartheta + \Phi_{1s}r\sin\vartheta$ in Cartesian and cylindrical coordinates, Φ_{1c} and Φ_{1s} being constants. If a finite number of longitudinal electrodes with a $\cos\vartheta$ or a $\sin\vartheta$ potential distribution are positioned on a cylinder of circular cross-section and radius $r = a$, an approximately homogeneous transverse field can be generated in the x- and/or y-directions. The degree of homogeneity depends on the number of electrodes, i.e. in effect the number of points used to approximate $\sin\vartheta$ or $\cos\vartheta$. By applying the potentials shown for the eight-pole deflector in Fig. 2C, each multipole coefficient can be calculated if the potential at the boundary $r = a$ is subject to Fourier analysis. To do this, a linear decrease in potential between each of the electrodes is assumed. If the gap width between the electrodes is described in terms of the angle ε, the following results can be derived for the x-component of the deflection field (the

y-component is derived analogously):

$$\Phi_{1c} = \frac{8}{\pi}\sqrt{2-\sqrt{2}}\frac{\sin(\varepsilon/2)}{\varepsilon}\frac{U_x}{a} \tag{34a}$$

$$\Phi_{3c} = \Phi_{5c} = 0 \qquad \Phi_{2\nu,c} = 0 \quad \text{for } \nu = 0, 1, 2, 3\ldots \tag{34b}$$

$$\Phi_{7c} = \frac{8}{49\pi}\sqrt{2-\sqrt{2}}\frac{\sin(7\varepsilon/2)}{\varepsilon}\frac{U_x}{a^7} \tag{34c}$$

The factor before U_x/a in Eq. (34a) is ≈ 0.97 when $0 \leqslant \varepsilon \leqslant \pi/8$. As Φ_{7c} is the first non-vanishing multipole coefficient, the deviation of the potential from linearity is proportional to r^7.

Before we make the transition to electrostatic multipole deflectors of finite length, it should be mentioned that the potentials applied by Kelly (1977) to the eight-pole deflector can be transformed into those used for the eight-pole deflector shown in Fig. 2C by rotating the coordinate system through $\pi/8$. Only two voltages (four potentials $\pm U_x$, $\pm U_y$) are required in both cases if a suitable resistance bridge is used (see Kelly, 1977 or Hutter, 1974).

Provided that symmetry is not destroyed, which multipole components vanish and which do not makes no difference to the results, irrespective of the length (finite or infinite) of the multipole deflector. Symmetry is not destroyed when the multipole deflector of finite length is free-standing or when it is sandwiched between electrodes with rotational symmetry (e.g. apertures or cylinders) whose potentials are equal to the average potential of the electrodes in the multipole deflector. The degree of homogeneity of the deflection field, however, can be degraded considerably under certain circumstances, as the deflection field along the z-axis, i.e. the dipole function $\Phi_1(z)$ is no longer constant; the following equation applies as a result of Eq. (5),

$$\begin{aligned}\varphi_1 = \varphi_1(r, \vartheta, z) = & \left\{\Phi_{1c}(z)r - \frac{1}{8}\Phi''_{1c}(z)r^3 + \cdots\right\}\cos\vartheta \\ & + \left\{\Phi_{1s}(z)r - \frac{1}{8}\Phi''_{1s}(z)r^3 + \cdots\right\}\sin\vartheta\end{aligned} \tag{35}$$

Eq. (35) shows that when we are dealing with multipole elements of finite length, terms involving r^3 occur in the series expansion of the potential (r^2 in the series expansion of the transverse field). Whether the transverse inhomogeneity is more strongly determined by the z-dependence of the dipole field or by the first non-vanishing higher-order multipole cannot as yet be decided; it is strongly dependent on the ratio a/l, where a is the inner radius of the boundary and l is the length of the electrodes in the multipole deflector.

Jumping ahead for a moment, it should be mentioned that only Φ_1, Φ_1'', and Φ_3 influence the third-order image aberration, if the electrostatic multipole deflector is perfectly aligned. It may now be wrong to assume that the eight-pole deflector with $\Phi_3 = 0$ and with $\Phi_7 \neq 0$ as the first non-vanishing coefficient is therefore good enough–even if the eight-pole element is relatively long. This is because when the multipole deflector is misaligned there exist, with respect to the misaligned optical axis, spurious dipole, quadrupole, hexapole and further multipole fields of higher orders, which are related to $\Phi_1(z)$ and the non-vanishing higher multipole functions $\Phi_\nu(z)$ ($\nu \geqslant 7$ for the eight-pole deflector) of the perfectly adjusted multipole deflector. It may therefore be reasonable to use more than eight electrodes, even if l is considerably larger than a. An example illustrating this would be the 20-pole deflector designed by Weidlich (1990). In this case, $l = 110$ mm and $a = 15$ mm, where it must be said that numerical methods were used. It is worth mentioning that for this 20-pole deflector, and similarly for the 20-pole deflectors previously referred to (Hutter, 1974; Idesawa, Soma, Goto, & Sasaki, 1983), the $\cos\vartheta$ boundary potential distribution for the x-deflection was generated by using cylindrical segment electrodes of different sizes and only potentials $\pm U_x$ were applied to these electrodes.

Analytical field calculations for multipole deflectors of finite length are hardly ever found in the literature. It is probable that closed-form solutions are only possible in a few cases, e.g. the sandwiched eight-pole deflector shown in Fig. 3. It is placed between two tubes with the same inner radius, a. If we assume that there is also a linear potential distribution in the axial gaps between the eight electrodes of the deflector and the grounded end tubes, this gives rise to a Dirichlet problem that can be described in terms of a Fourier–Bessel series (see Eq. (11)). The constants in Eq. (11) are related to the boundary potentials. For $\nu = 1$ and $r = 0$, the axial distribution for the deflection field strength $E_x(0, 0, z) = -\Phi_{1c}(z)$ is obtained; this has been derived by Plies and Elstner (1989b). For a sandwiched multipole deflector made from eight solid, cylindrical electrodes, Kunze and Jansen (1987, 1989) have derived the spatial potential in analytical form. The centers of the cylinders were arranged regularly in a circle about the optical axis. In the first publication, the multipole deflector was placed between two grounded planes; in the second, the authors opened the grounded planes with two coaxial grounded tubes of semi-infinite length.

For a very long multipole deflector, $\Phi_{1c}(z)$ can be described in terms of a uniform rectangular ("top-hat") function, while $\Phi_{1c}(z)$ is bell shaped for a deflector of finite length. Grümm and Spurny (1956) have calculated the paraxial trajectory and the deflection image aberrations in closed form for the bell-shaped model field $\Phi_{1c} = \Phi_{1c}(0)/\cosh^2(z/h)$. Lenc and Lencová (1988) have stated the third-order deflection aberrations for the "top-hat field". They also discussed earlier work on this field. Various assumptions were made about the hexapole component of the field. They also take the refraction of the rays in the sharp cut-off fringing field into account.

Magnetic Deflectors. Toroidal or saddle coils are the most frequently used type of magnetic beam deflector. If there is no ferromagnetic material in the vicinity, the magnetic field produced by the short deflector can usually be given in an analytical form, which is not the case for electrostatic deflectors. The calculations are based on the Biot–Savart law, the integral form of the vector potential **A** or the integral form of the scalar magnetic potential ψ. They are evaluated for a line current I in a given conductor. Because of the circular arcs encountered at the ends of the saddle coil, its analysis involves elliptic functions (e.g. see Urankar, 1980 or Munack, 1990). This means that extensive numerical evaluation is required. If, on the other hand, the whole field over space is not of prime interest, an expansion using powers of r in the integrands of the integral formulae provides the lowest-order multipole components $\Psi_\nu(z)$ in the form of simple functions. The dipole function $\Psi_1(z)$ and the hexapole function $\Psi_3(z)$ for both toroidal and saddle coils were obtained by Munro and Chu (1982a) in this manner. They considered the simple but important case of conductors on a cylindrical surface. Lencová, Lenc, and van der Mast (1989) state an analytical formula for $\Psi_1(z)$ in the case of a tapered saddle coil.

We shall now consider the construction of multiconductor coils using a saddle coil on a cylindrical former as an example. The following equations apply to a saddle coil with a single (double winding) as shown in Fig. 5:

$$\Psi_{2\nu,c} = 0 \quad \Psi_{2\nu+1,c} = \mu_0 I \frac{\sin[(2\nu+1)\Theta]}{2\nu+1} F_{2\nu+1}(z,l,a) \quad \text{for } \nu = 0,1,2,3,\ldots \tag{36a}$$

$$\Psi_{1c} = \mu_0 I \sin\Theta F_1(z,l,a) = \mu_0 \frac{I}{\pi a} \sin\Theta \left\{ \frac{z+l/2}{a}(f_1 + f_1^3) - \frac{z-l/2}{a}(f_2 + f_2^3) \right\} \tag{36b}$$

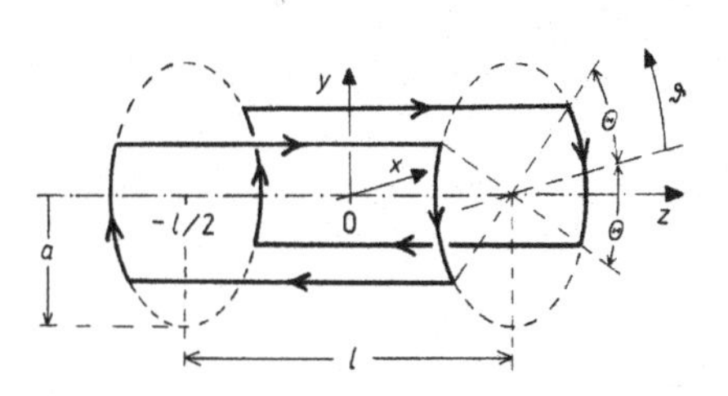

Figure 5 Geometry of a simple saddle coil consisting of a (double) current-carrying winding arranged on a cylindrical surface.

$$\Psi_{3c} = \mu_0 I \frac{\sin 3\Theta}{3} F_3(z, l, a) = \frac{\mu_0 I}{24\pi a^3} \sin 3\Theta \left\{ \frac{z + l/2}{a} (8f_1 + 4f_1^3 + 3f_1^5 + 5f_1^7) \right.$$
$$\left. - \frac{z - l/2}{a} (8f_2 + 4f_2^3 + 3f_2^5 + 5f_2^7) \right\} \tag{36c}$$

and

$$f_1 = \frac{a}{[(z + l/2)^2 + a^2]^{1/2}} \qquad f_2 = \frac{a}{[(z - l/2)^2 + a^2]^{1/2}} \tag{36d}$$

In these equations Θ is the semi-angle of the circular arcs, a their radius, and l the length of the coil. Eqs. (36b) and (36c) can be transformed into Eqs. (36) and (37), which were given in the paper by Munro and Chu (1982a). Munro and Chu obtained their formulae by using the method of scalar magnetic potentials (magnetic sheath produced by a layer of magnetic dipoles which are equivalents of the current; e.g. see Durand, 1968). This method proved to be very elegant and convenient in this case. Eqs. (36b) and (36c) were obtained from the Biot–Savart law. This has the advantage that the calculation tells us which summands are related to the four straight conductors and which are due to the four arcs of a circle. Eq. (36b) contains the cubes of the functions f_1 and f_2, while Eq. (36c) contains the seventh powers; these terms are due to the arcs of a circle. If $a \ll 1$, the following equations are valid within the saddle coil

$$\Psi_{1c} \approx \mu_0 \frac{2I}{\pi a} \sin \Theta \quad \text{and} \quad \Psi_{3c} \approx \mu_0 \frac{2I}{3\pi a^3} \sin 3\Theta \tag{37}$$

As can be seen from Eq. (36c), the hexapole component of the saddle coil with a current loop pair vanishes at $\Theta = \pi/3 = 60°$, irrespective of the ratio a/l. For two current loop pairs, which are positioned at $\Theta_1 = 42°$ and $\Theta_2 = 78°$ and which have the same length l and the same radius a, Ψ_{3c}, and Ψ_{5c} vanish (Munro, 1980).

If a very large number N of very thin conductors, each carrying a current I, were to be positioned on the surface of a cylinder so that their distribution was proportional to $\sin\vartheta$ relative to the azimuth ϑ, the following equation would describe the odd multipole components:

$$\begin{aligned}\Psi_{2\nu+1,c} &= \mu_0 \frac{NI}{2\nu+1}\int_0^{\pi/2} \sin\big[(2\nu+1)\vartheta\big] F_{2\nu+1}(z,l,a)\sin\vartheta\,d\vartheta \\ &= \mu_0 \frac{NI}{2\nu+1} F_{2\nu+1}(z,l,a)\int_0^{\pi/2} \sin\big[(2\nu+1)\vartheta\big]\sin\vartheta\,d\vartheta \\ &= \begin{cases} \mu_0 NI\frac{\pi}{4}F_1(z,\ l,\ a) & \text{for } \nu=0 \\ 0 & \text{for } \nu \geqslant 1 \end{cases}\end{aligned} \tag{38}$$

In other words, all the multipole components except the dipole component would vanish, as the even multipole components are all zero because of the symmetry of the arrangement. In practice, using a multiconductor saddle coil, this arrangement can only be approximated. The main reasons for this are the finite number of conductors used to approximate $\sin\vartheta$ and the topology of the arrangement, which means that a and l are not the same for all windings. This implies that the radius a and/or the length l are functions of ϑ, which would make it improper to place the factor $F(z, l(\vartheta), a(\vartheta))$, before the integral. If wire windings of the same length l are used, the radius a is different for different windings. In the case of a printed saddle coil (flexible, etched copper laminate wound on a cylindrical former; e.g. see Herrmann, Menadue, and Pearce-Percy, 1976), the radius a of all circular conductors is the same, but the length l of each of the windings is different. If a printed saddle coil of this kind is used, even the hexapole component which influences the image aberrations is not identically equal to zero. (In the case of a toroidal coil, there is a second cylindrical surface available for guiding the windings. This means that the topological restrictions on generating a pure dipole field are not as severe.)

In the case of a real multiturn saddle coil, when the number N of the current loop pairs is large and l and a are not the same for any current loop pair, the total field is obtained by summing over all the terms in Eq. (36). It is always possible to find an arrangement of this kind where $|\Psi_3(z)|$ remains below a limit and the integral over $\Psi_3(z)$ vanishes. If the stricter condition $\Psi_3(z) \equiv 0$ is to be fulfilled, Ohiwa (1977, 1978) suggests using a correction coil in addition to the basic deflection coil. The correction coil generates an additional hexapole field. According to Ohiwa, this correction coil can be excited in conjunction with the basic coil. Mills and Morgan (1972) published a flux theorem for the design of magnetic coil ends and found a winding of a dipole saddle coil end, lying in a circular cylinder, which develops no higher harmonics.

Further information on designing saddle and toroidal coils and on performing analytical field calculations can be found, for example, in publications by Haantjes and Lubben (1957, 1959), Kramer (1967), Kaashoek (1968), Ginsberg and Melchner (1970), Dekkers (1974), Sheppard and Ahmed (1976), and Hanssum (1984, 1985, 1986). Kaashoek has studied complicated coils of the type found in television tubes. Some of the authors mentioned above do not take the finite length of the coil into account, others do not consider the coil current that flows in a direction perpendicular to the optical axis. These assumptions are often justified but need not be so in general. Near the axis, not all the authors use the expansion of the field with planar multipoles or the Fourier expansion, but instead use Cartesian polynomials. The following two formulae (for a deflection field in the x-direction) are useful for making transformations:

$$\begin{aligned} B_0 &= B_x(0,0,z) = -\Psi_{1c}, \\ B_2 &= \frac{1}{2}\frac{\partial^2 B_x}{\partial y^2}\bigg|_{x=y=0} = \frac{1}{2}\frac{\partial^2 B_y}{\partial x \partial y}\bigg|_{x=y=0} = 3\Psi_{3c} + \frac{1}{8}\Psi_{1c}'' \end{aligned} \tag{39}$$

For the long coil, $\Psi_{1c}(z)$ is shaped almost like a box, whereas the function is bell shaped for the short coil. Apart from the early work done on the subject, the magnetic "top-hat model field" was last used successfully by Lenc and Lencová (1988) and Lencová (1988a). Lencová (1988a) also states the image aberration coefficients for a magnetic cosine field (a cosine period increased by the amplitude). Grümm and Spurny's (1956) bell-shaped model field, already mentioned in conjunction with the electrostatic deflectors, can also be used for magnetic deflection fields. Its image aberration coefficients are given in the work of Grümm and Spurny (1956).

Some Special Fields

In the vicinity of the cathode tip of a field emission source, at least, analytical solutions to Laplace's equation must be used. The sphere on orthogonal cone model by Dyke, Trolan, Dolan, and Barnes (1953) is particularly well known and was previously frequently used. It can, however, only be adapted to the boundary conditions on the cathode surface and not to other electrodes positioned further away. Kern (1978) has extended this model and was able to satisfy all the boundary conditions by using analytical field calculations near the tip and numerical FDM calculations in the outer regions. Jumping ahead for a moment to numerical field calculations, it should be said that potential problems of this kind, where the size of the electrode features varies greatly (four to five orders of magnitude difference between the dimension of the cathode tip and that of the electrode elsewhere) are best solved today with the CSM, as described by Hoch, Kasper, and Kern (1978). The success of the CSM is due to its analytical character. The example from Kasper (1979), shown in Fig. 9, uses plane, infinitely thin circular apertures, charged rings, a homogeneously charged cylinder (of radius R), an infinitely thin wire (located on the axis and consisting of two sections with constant but different charges per unit length) and a point charge (Q_s) at the end of the thin wire to model a field emission device.

Grids are usually used for retarding-field spectrometers. The energy resolution of a retarding-field spectrometer depends strongly on the penetration factor of the retarding field at the retarding grid (e.g. see Simpson, 1961 or Menzel and Kubalek, 1983). Analytical formulae for the penetration factor and the field between two plane grids can be found in Ollendorf (1932), Küpfmüller (1965), and Huchital and Rigden (1972), for example. It should be noted that in each case the grid only comprised a row of infinitely long, straight wires. Two-dimensional problems of this kind can often be solved using methods from the theory of analytical functions. Lists of examples using conformal mapping have been compiled by Koppenfels and Stallmann (1959), Betz (1964), and Binns and Lawrenson (1973). Resorting to analytical methods of field calculation can prove to be very helpful and useful for estimating critical field strengths along edges or arranging slot apertures to limit stray fields from deflection magnets.

In the fields of e-beam testing and electron optical inspection, it is not just the fields of the electron optical elements that play a role, the fields of the probe are important too. Above and between the interconnections of an IC, there exists a local microfield which influences at least the SEs released by the PEs (Nakamae, Fujioka, & Ura, 1981), sometimes even the low-voltage PE probe (Ura, 1981; Ishii, Hyozo, Tada, & Maki, 1989). Ura, Fujioka, and Yokobayashi (1980) and Nakamae et al. (1981) give a relatively simple formula for the microfield of a simple arrangement of parallel interconnections. The formula was successfully used for fast calculations of the SE trajectories in the microfield. Some caution should be used, however, when the height of the interconnections is no longer significantly less than their width. The formula from Ura et al. (1980) does not then take the height of the interconnections into account, as the authors themselves state.

During wafer inspection, it is desirable to perform in-line dimensional measurements on the trench cells of DRAMS (dynamic random access memories) and to detect defects in the trench cells non-destructively. This can be achieved today only by destructive cross-sectioning. If electrostatic extraction of the SEs from the trench is applied, the penetration of the extraction field into the trench cell can be estimated analytically. This is because the wall of the empty trench cell is made from highly doped polysilicon (specific resistance is only two orders of magnitude worse than aluminum) and so can be modeled as a tubular electrode. As with the two-tube lens, the use of separation of variables leads to a Fourier–Bessel series for the potential.

2.2 Numerical Methods

The three most important methods for calculating the fields of electron optical elements are the FDM, the FEM and the class of integral equation methods (IEMs). The boundary element method (BEM) belongs to the last class, which is also referred to as the surface charge density method (CDM) in electrostatic applications. Closely related to the CDM is the CSM already mentioned. With this method, equivalent charges are arranged *inside* the electrode to generate the boundary potential required at its surface.

We will discuss the FDM, FEM, CDM and CSM briefly in the order given and illustrate them with a few examples that are relevant to e-beam testing or electron optical inspection. The textbook by Hawkes and Kasper (1989a) contains an exhaustive overview of methods for numerical calculation of fields. There are also earlier overviews, for example those by Weber (1967), Mulvey and Wallington (1973), Bonjour (1980), Kasper (1982), and Kasper (1984a). The work of Becker (1989, 1990) and Kasper and Ströer (1989, 1990) must be considered to be a significant advance in the numerical calculation of magnetic fields (of lenses, quadrupoles, deflection systems and bending magnets). The methods described in the second publication cited in each case can now also be applied to systems containing yokes with finite permeability. Also worthy of mention are the numerical field calculations for true three-dimensional problems carried out by Rouse and Munro (1989, 1990).

It is also interesting to note that field calculation programs for electron optical systems are now officially available for purchase from the Delft Particle Optics Foundation and Imperial College, London. General-purpose FDM, FEM or BEM programs, which are marketed by many companies, are often satisfactory for some problems. However, general-purpose problem solvers of this kind do much to confirm the general validity of Murphy's law by being unable to deal with special cases of interest. Important factors when assessing commercial programs are the availability of pre-processors (e.g. for generating the FEM lattice) and of post-processors (e.g. for processing the field results for ray tracing).

Even today, it still takes a lot of computer time to handle true three-dimensional problems which have a complicated boundary and a large basic region. With a few three-dimensional problems, however, it is possible to express the three-dimensional potential distribution as a sum of two-dimensional potentials. The following type of Fourier expansion using cylindrical coordinates (Schwertfeger & Kasper, 1974; Kasper, 1976; Kasper & Lenz, 1980; Kasper, 1981) has proved useful for quadrupole lenses or deflection elements (e.g. electrostatic eight-pole deflectors or the magnetic toroidal coil):

$$\varphi = \varphi(z, r, \vartheta) = \sum_{\nu=0}^{\infty} \varphi_\nu = \sum_{\nu=0}^{\infty} \mathrm{Re}\{P_\nu(z, r)\mathrm{e}^{\mathrm{i}\nu\vartheta}\} \tag{40a}$$

$$= \sum_{\nu=0}^{\infty} r^\nu \mathrm{Re}\{p_\nu(z, r)\mathrm{e}^{\mathrm{i}\nu\vartheta}\} \tag{40b}$$

If Eqs. (40a) and (40b) are substituted into Laplace's equation expressed in terms of cylindrical coordinates, two-dimensional decoupled partial differential equations (PDEs) are obtained for the functions $P_\nu(z, r)$ and $p_\nu(z, r)$:

$$\frac{\partial^2 P_\nu}{\partial r^2} + \frac{1}{r}\frac{\partial P_\nu}{\partial r} + \frac{\partial^2 P_\nu}{\partial z^2} - \frac{\nu^2}{r^2} P_\nu = 0 \tag{41a}$$

$$\frac{\partial^2 p_\nu}{\partial r^2} + \frac{2\nu + 1}{r}\frac{\partial p_\nu}{\partial r} + \frac{\partial^2 p_\nu}{\partial z^2} = 0 \tag{41b}$$

The boundary conditions which must be satisfied by the Fourier coefficients $P_\nu(z, r)$ and $p_\nu(z, r)$ are obtained by means of the Fourier inverse transform on Eqs. (40a) and (40b) from the potential $\varphi(z, a(z))$ on a surface $r = a(z)$ with rotational symmetry. The advantage of the Fourier expansion given by Eqs. (40a) and (40b) is the reduction in the number of dimensions that have to be considered; instead of the three-dimensional Laplace

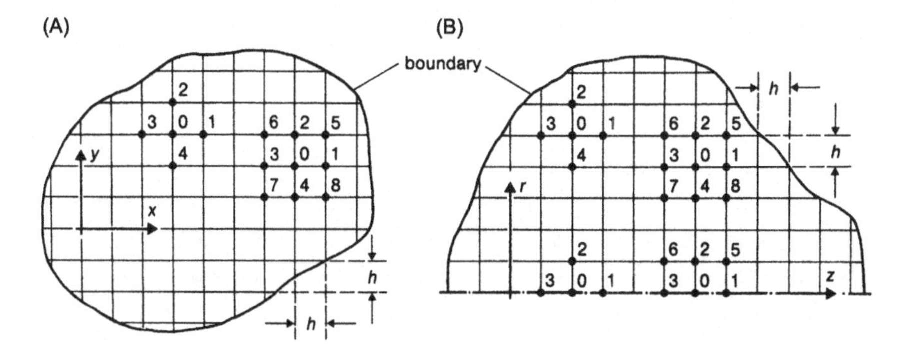

Figure 6 Configurations and notation used for the finite difference formulae of regular internal and regular on-axis nodes in a square mesh grid. (A) Planar Cartesian coordinates, (B) cylindrical coordinates.

equation only a couple of two-dimensional PDEs need be solved. This can be accomplished using the FDM, for example. Two-dimensional PDEs similar to those in Eqs. (41a) and (41b) can also be obtained for the three-dimensional Poisson equation or the PDE describing the magnetic vector potential. To do this, however, the Fourier expansions of the space charge density or the current must be used.

The FDM

The FDM is a well-known discrete method for solving PDEs numerically. As it was introduced by Liebmann (1918) it is sometimes referred to as "Liebmann's method". The basic idea behind it is to divide the whole region of the boundary value problem (Dirichlet's problem, Neumann's problem or the mixed problem) up by means of a rectangular or square lattice (Fig. 6), and then to determine the potentials at the points on the lattice from the potentials on the boundary. When solving Laplace's equation in this way, i.e. determining the potentials at the internal lattice points, the partial derivatives in Laplace's equation are replaced by finite difference approximations. In this way, equations are set up that relate the potential at a certain point to the potentials at neighboring points.

In the two-dimensional case (Fig. 6A), using a square lattice with width h, the five-point formula or second-order approximation is obtained for a regular point inside the domain of solution

$$4\varphi_0 = \varphi_1 + \varphi_2 + \varphi_3 + \varphi_4 + 0\left(h^4\right) \tag{42}$$

as is the nine-point formula or fourth-order approximation

$$20\varphi_0 = 4(\varphi_1 + \varphi_2 + \varphi_3 + \varphi_4) + \varphi_5 + \varphi_6 + \varphi_7 + \varphi_8 + 0(h^6) \tag{43}$$

The order of the discretization error is included in both formulae (the indices refer to the numbering of the lattice points in Fig. 6A and not the Fourier components in, say, Eq. (40a)). When rotational symmetry exists (Fig. 6B), the following five-and nine-point formulae are obtained for a regular point inside the region:

$$8\varphi_0 = 2(\varphi_1 + \varphi_2 + \varphi_3 + \varphi_4) + \frac{h}{r_0}(\varphi_2 - \varphi_4) + 0(h^4) \tag{44}$$

and

$$\left.\begin{aligned} & c_0\varphi_0 = \sum_{i=1}^{8} c_i\varphi_i + 0(h^6) \\ & \text{with} \\ & c_0 = 120m(8m^2+3) \quad m = h/r_0 \\ & c_1 = c_3 = 12m(16m^2+7) \\ & c_2 = 2(96m^3 + 48m^2 + 30m + 23) \\ & c_4 = 2(96m^3 - 48m^2 + 30m - 23) \\ & c_5 = c_6 = 48m^3 + 24m^2 + 18m + 13 \\ & c_7 = c_8 = 48m^3 - 24m^2 + 18m - 13 \end{aligned}\right\} \tag{45}$$

The following formulae are obtained for a regular point on the axis:

$$6\varphi_0 = 4\varphi_2 + \varphi_1 + \varphi_3 + 0(h^4) \tag{46}$$

and

$$58\varphi_0 = 34\varphi_2 + 5(\varphi_1 + \varphi_3) + 7(\varphi_5 + \varphi_6) + 0(h^6) \tag{47}$$

The derivation of formulae (42)–(47) can be found in, for example, Durand (1964, 1966a, 1966b). These references also contain formulae for rectangular lattices, for various kinds of points near the boundary (i.e. irregular internal points) and formulae of the second to sixth orders for the three-dimensional Cartesian case.

Kasper has also published five- and nine-point formulae for PDEs of the same type as Eq. (41b) (Kasper, 1976, 1982, 1984b, 1990). Radially increasing grids are recommended for calculating potential distributions in electron guns with tip cathodes. This procedure keeps the number of points within manageable limits. A "spherical coordinates with increasing mesh" (SCWIM) grid was proposed by Kang, Orloff, Swanson, and Tuggle (1981), Kang, Tuggle, and Swanson (1983) in connection with a five-point discretization formula. Killes (1985) worked out a similar "spherical coordinates with exponentially increasing mesh" (SWEM) grid using Kasper's nine-point formula (Kasper, 1984b).

If the five- or nine-point formulae are written out for each of the N ($= 10^3$–10^4) lattice points, a system of N linear equations is obtained. It is scarcely feasible to solve this system of equations by direct methods. Usually iterative methods are used to solve systems of finite difference equations whose coefficient matrix is sparse. The successive over-relaxation method (SOR) is often employed; it is extensively discussed in the mathematical literature as well as in articles on numerical field computation (e.g. Weber, 1967 or Binns & Lawrenson, 1973). A further development of SOR is the well-known successive line over-relaxation method (SLOR), where the potentials are not iterated point by point, but rather line by line. Kasper and Lenz (1980) compared the SOR and SLOR, coming to the conclusion that the SLOR is more stable and converges slightly faster than the SOR.

Fig. 7 shows the equipotential lines in three orthogonal planes through the center of a trough in a block of insulating material which has a constant negative charge density. This example, originating from Rouse and Munro (1990) is important for simulating electron optical line width measurements on insulating specimens. To perform this field calculation, Rouse and Munro (1990) chose the FDM using seven-point difference equations on a three-dimensional Cartesian grid and a variable spacing on each of the three coordinate axes, thus allowing a greater grid point density to be used in critical regions. This three-dimensional field solver designed by Rouse and Munro (1989) was successfully used by the author to simulate the electrostatic field in the upper section of the spectrometer (region between the retarding-field grid and detectors) in the retarding-field spectrometer of an EBT described by Frosien and Plies (1987).

The FEM

The FEM was proposed by Courant (1943). After the development of more powerful computers, this method could be used in a number of practical applications, the first being civil engineering (Zienkiewicz and Cheung, 1965). The FEM was first used in electron optics by Munro (1970, 1971) to calculate the fields of magnetic and electrostatic round lenses.

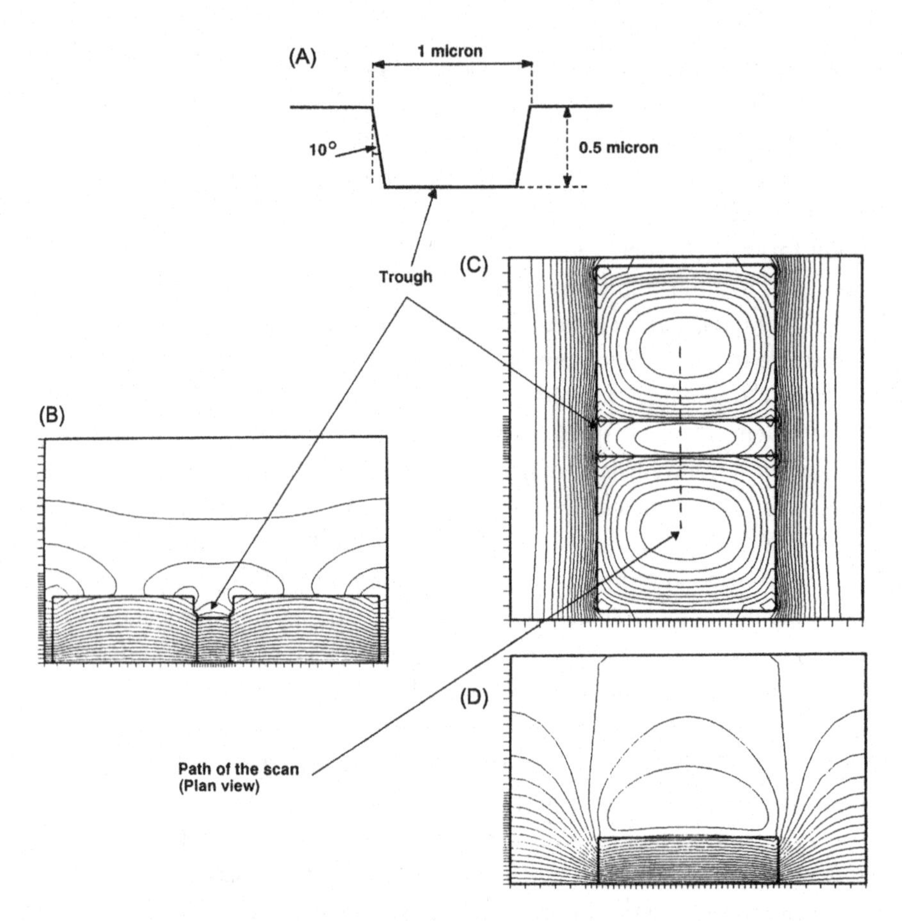

Figure 7 (A) Geometry of a trough in a block of insulating material. (B)–(D) Equipotentials in three orthogonal planes through the center of the trough in the block to which a constant negative surface charge density is applied. *From Rouse and Munro (1990).*

With the FEM, we do not solve the PDE but instead the associated variational problem. The solution domain is subdivided using a mesh of small, irregular quadrilaterals. Most often, the quadrilaterals are further subdivided into triangular finite elements. It is (usually) assumed that the scalar or vector potential is a linear function over the triangular finite elements. This means that the potential throughout each finite element is uniquely determined by the potential at the nodes of each finite element. The potentials at these points are then included in the contributions of a finite element to the functional. The variation of the functional then gives a set of algebraic equations which relate the potential at a particular point to the potentials of the neighboring points. The boundary conditions are inserted into these finite element equations.

Unlike the FDM, the FEM can be used to make all the boundaries of electrodes or pole pieces coincide on mesh lines. Moreover, the complete magnetic circuit including coil windings can be analyzed and the finite permeability $\mu = \mu_0 \mu_r$ can also be taken into account as a function of location. A list comprising a few relevant PDEs and functionals F for the magnetic case, as cited by Munro (1973), will now be given:

(i) No currents in the region of interest, and the magnetic material is unsaturated

$$\mathbf{B} = -\nabla \psi \qquad \Delta \psi = 0 \tag{48a}$$

$$F = \iiint\limits_{\text{volume}} \frac{1}{2\mu} (\nabla \psi)^2 \mathrm{d}v \tag{48b}$$

(ii) Currents included ($\mathbf{j}$ = current density), and the magnetic material is unsaturated

$$\mathbf{B} = \nabla \times \mathbf{A} \qquad \nabla \times \left(\frac{1}{\mu} \nabla \times \mathbf{A} \right) = \mathbf{j} \tag{49a}$$

$$F = \iiint\limits_{\text{volume}} \left\{ \frac{1}{2\mu} (\nabla \times \mathbf{A})^2 - \mathbf{j} \cdot \mathbf{A} \right\} \mathrm{d}v \tag{49b}$$

(iii) Currents and saturation effects are included:
PDE as Eq. (49a) and

$$F = \iiint\limits_{\text{volume}} \left\{ \int_0^B H \mathrm{d}B - \mathbf{j} \cdot \mathbf{A} \right\} \mathrm{d}v \tag{50}$$

In the case of axis-symmetric magnetic lenses, the volume integrals of Eqs. (48b), (49b), and (50) are reduced to area integrals. The formulae can be found in Munro's article (1973). When A_ϑ for a magnetic lens has been calculated, the lines of magnetic flux are obtained from rA_ϑ = constant. To solve electrostatic problems (with no space charge) $\mathbf{E}$ should be substituted in Eqs. (48a) and (48b) for $\mathbf{B}$, φ for ψ and the permittivity $\varepsilon = \varepsilon_0 \varepsilon_r$ for the permeability μ.

In the case of unsaturated magnetic material, the set of finite element equations is linear and can be solved by means of standard methods. Usually direct methods of solution are used, e.g. inverting the matrix defining the system of linear equations. Iterative methods such as the SOR and SLOR, which have been mentioned previously, can also be used. Lencová and Lenc (1986) have developed a very fast iterative procedure involving the use of the preconditioned conjugate gradient method.

At saturation, i.e. when the relative permeability $\mu_r = \mu/\mu_0$ is not $\gg 1$, the system of finite element equations is not linear. This makes the use of iterative methods unavoidable. Munro (1973) combined direct techniques with Newton's method to solve this system of equations.

Fig. 8 shows the results of an FEM computation and also the grid used. This example refers to the variable axis immersion lens (VAIL) as used by Dubbeldam (1989) in a laboratory EBT for focusing the PEs and collimating the SEs. Fig. 8A shows the geometry of the various ferromagnetic parts (iron circuit, pole piece and mirror plate) and the shape of the two coils used (excitation coil and auxiliary coil for generating the slow left-field decay). In Fig. 8B, the FEM grid used was additionally superposed. Fig. 8C shows the flux density lines and the axial curve $B(z)$. In addition, the magnitude of the flux density in the ferromagnetic parts is shown as a gray scale. Lencová, who performed the FEM computation of this example, used Neumann boundary conditions at the artificial outer boundaries.

The FEM is, of course, not restricted to field calculations with rotational symmetry, it can also be used to calculate deflection fields, e.g. when the Biot–Savart law is not applicable. This is the case when deflection yokes made from ferromagnetic material are used, or when the deflection coils are close to the magnetic iron circuit of the lens. Examples of this kind dealing with saddle coils and toroidal coils are given by Munro and Chu (1982a) and Lencová et al. (1989) amongst others.

A crucial disadvantage of the FEM is that we cannot extend the grid in space without limit. The error that arises when artificial boundaries are introduced is particularly great when the fringing field has not decayed sufficiently. This situation is often encountered with open electrode or pole piece structures, e.g. with Mulvey's single pole piece lens (Mulvey, 1975, 1982, 1984). Mulvey and Nasr (1981) have obtained considerably better results with the selected intermediate boundary method which they proposed. This method uses several different boundaries and runs a sequence of three computations so that the results from the previous runs for larger boundaries supply boundary values for the smaller (intermediate) ones. In their publication, Mulvey and Nasr have also proposed the differential-integral method for checking, smoothing and improving the axial flux density distribution $B(z)$ by calculating the contributions due

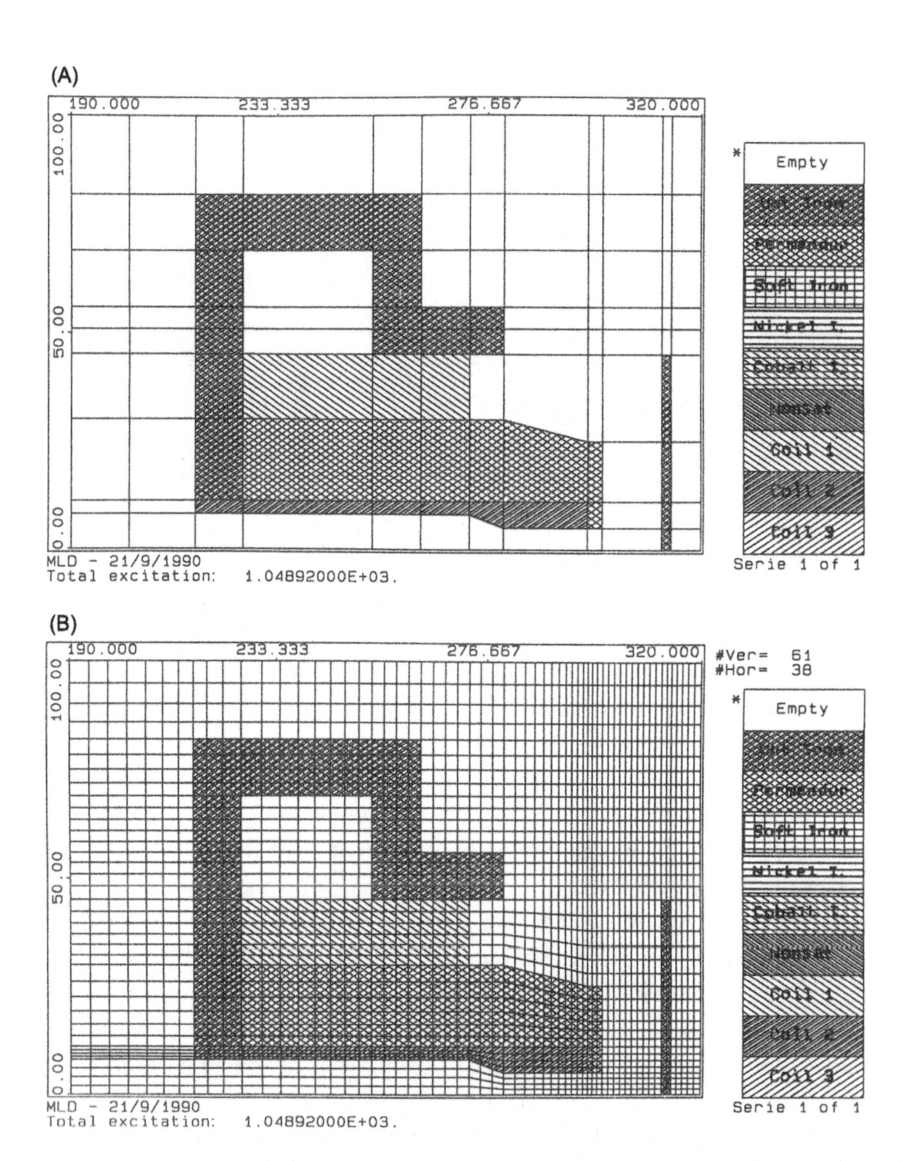

Figure 8 FEM computation of a variable axis immersion lens as used by Dubbeldam (1989) in a laboratory EBT. (A) Geometry and position of the various ferromagnetic parts and the two coils. (B) As (A) plus superposed FEM grid. (C) Results: flux density lines, axial distribution $B(z)$ of the flux density and representation of the strength of the flux density in the ferromagnetic parts with the aid of a gray scale. In part (C) the coils have been indicated by large crosses in the usual way. The keys to the right of (A) and (B) are, from top to bottom: Empty, Upt Iron (Czechoslovakian trade mark), Permendur, Soft Iron, Nickel I. (I = iron), Cobalt I., Nonsat (nonsaturated), Coil 1, Coil 2, Coil 3. *Courtesy of B. Lencová.*

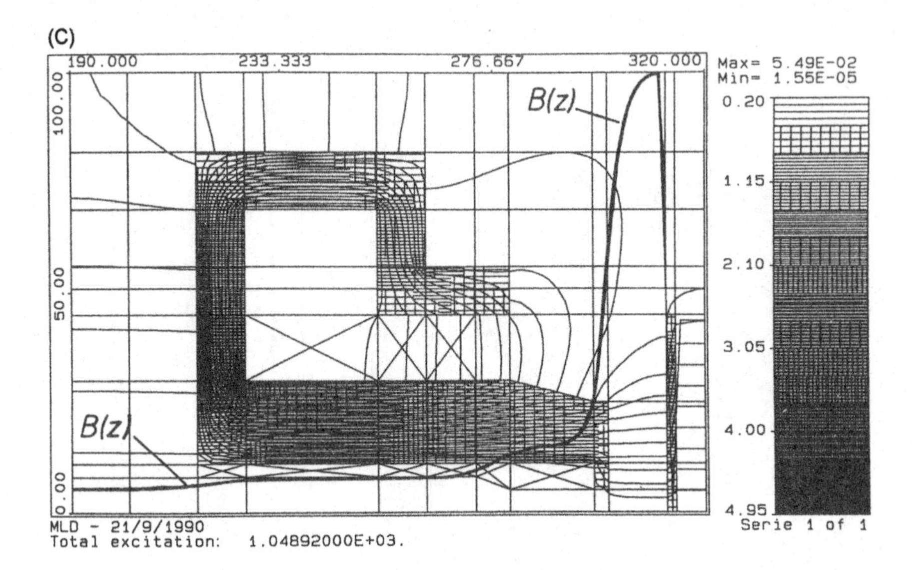

Figure 8 (*continued*).

to the coil and the magnetic circuit. The disadvantage of the FEM encountered with magnetic lenses having an open magnetic circuit has been eliminated by Lencová and Lenc (1982, 1984) by using additional infinite elements. The shape functions for the potential which is to be evaluated over infinite elements of this kind must be consistent with the asymptotic behavior of the actual potential.

Some problems with the FEM have been reported by several authors in the past; references are given by Lencová (1988b). The paper by Tahir and Mulvey (1990) reviews various errors that can occur in the course of an FEM computation (e.g. during grid generation) and presents certain useful diagnostic checks which can be incorporated into the FEM computer program to warn (inexperienced) operators when errors are about to occur. Kasper and Lenz (1980) and Kasper (1982) have criticized the use of a linear approximation for scalar potentials (Munro, 1973) and flux functions (Bonjour, 1980) when FEM computations are being made to solve lens problems. Kasper and Lenz (1980) showed that the FEM is less accurate than the FDM under these circumstances, at least in the paraxial domain. This disadvantage of the FEM can be overcome by using the higher-order approximations for the potential stated by Silvester and Konrad (1973), Munro (1987a), and Zhu and Munro (1989). Lencová (1988b) also mentions that more accurate on-axis results can be obtained by using special procedures for the finite element triangles with two vertices on the axis.

The CDM and the CSM

Although there is a very wide range of IEMs and BEMs to choose from, only the CDM (without surface dipole layer) and the CSM will be discussed in this section. This is because the CSM (a relative of the CDM) is particularly well suited for calculating the fields associated with geometries which have markedly different feature sizes. This kind of problem is encountered with, say, e-beam generators having pointed cathodes or with the calculation of microfields over the interconnections of an IC which is being operated in a macroscopic electrostatic field to extract SEs.

The physical idea behind the CDM is to find the surface charge distribution $\sigma(x, y, z)$ which is equivalent to the voltages applied to the electrodes. If σ has been found, the potential at any point in space (x_0, y_0, z_0) is given by the Coulomb surface integral

$$\varphi(x_0, y_0, z_0) = \frac{1}{4\pi\varepsilon_0} \iint\limits_{\text{surface}} \frac{\sigma(x, y, z)\mathrm{d}s(x, y, z)}{\sqrt{(x_0 - x)^2 + (y_0 - y)^2 + (z_0 - z)^2}} \tag{51}$$

If the boundary values are inserted into Eq. (51), the equation for φ, an integral equation for σ is obtained. Solving this equation is the most difficult aspect of the CDM, but an approximate solution can be found in the following manner. Each electrode is divided up into a finite number, N, of small subelectrodes each having a constant surface charge density. The potential of each subelectrode is taken to be the potential at its center. Proceeding on these assumptions, a system of linear algebraic equations for the surface densities σ_i of the subelectrodes is obtained. The right-hand side gives the potentials of the subelectrodes φ_j:

$$\sum_{i=1}^{N} P_{ij}\sigma_i = \varphi_j \quad j = 1, 2, 3, \ldots, N \tag{52a}$$

with

$$P_{ij} = \frac{1}{4\pi\varepsilon_0} \iint\limits_{s_i} \frac{\mathrm{d}s_i}{\sqrt{(x_i - x_j)^2 + (y_i - y_j)^2 + (z_i - z_j)^2}} \tag{52b}$$

The matrix elements P_{ij}, which depend only on the geometry, can often be obtained only by numerical integration over the surfaces s_i of the ith subelectrode. Caution has to be exercised when $i=j$ since this has to be handled as a special case to deal with the singularity involved.

Cruise (1963), Harrington (1968), Singer and Braun (1970), Rauh (1971), Read (1971), Read, Adams, and Soto-Montiel (1971), Adams and Read (1972), and Birtles, Mayo, and Bennet (1973) were all pioneers of this method, which is sometimes also referred to as the method of moments. The list of references does not claim to be complete, and in any case no attempt has been made to go beyond 1973. An important advantage of this method is that after determining the equivalent surface charges, the potential and field at any point in space is known and no further interpolation of any kind is required in order to calculate the electron trajectories numerically; this is not the case with the FDM and the FEM. Harting and Read (1976) have used the CDM to tabulate the cardinal elements and image aberrations of electrostatic lenses. Using the CDM, Munro and Chu (1982b) have calculated the field of electrostatic deflectors with cylindrical and/or conical segments. A wide range of examples could be cited to illustrate the applications of the method, but we will restrict ourselves to two–the calculation for a complex realistic electron gun (Andretta, Currado, Marini, & Zanarini, 1985) and the work of Ströer (1988). Ströer made use of triangular boundary elements carrying a linearly distributed surface charge and was able to iteratively calculate configurations of electrodes with rotational symmetry which were not necessarily aligned, e.g. electrostatic eight-pole deflectors and quadrupole lenses.

A major disadvantage of the CDM is the complicated evaluation of the improper integrations of the diagonal matrix elements P_{ii} in Eq. (52b). Hoch et al. (1978) have published a method which avoids these difficulties. There, the solution involves arranging charged rings and thin circular apertures *inside* an electrode with rotational symmetry so that the total effect of all the equivalent charges produces the applied voltage at the surface of the electrode. This is checked at a number of points on the surface of the electrode. Hoch et al. (1978) decided to use the number of charged rings as the number of test points. As all the rings and plane circular apertures were arranged *inside* the real electrodes, there are no field singularities in the space external to the surface of the electrode and on the electrode itself. Once the magnitude and position of the equivalent charge elements have been calculated iteratively, the potential and field for the entire space can be determined by superposition (for analytical solutions for rings and circular apertures, see Section 2.1). We will refer to this technique as the CSM, thus following Schönecker, Rose, and Spehr (1986, 1990). The version of

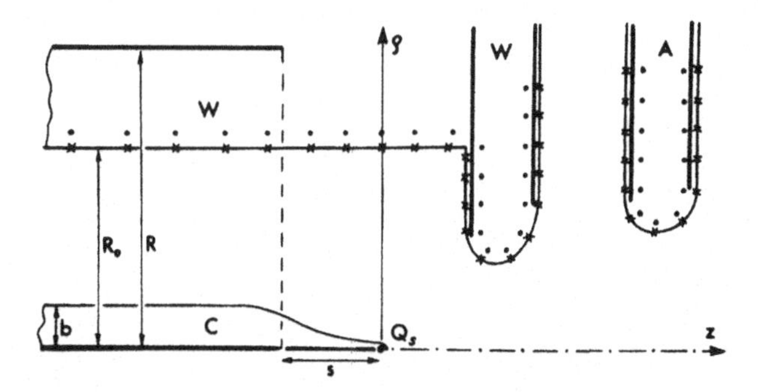

Figure 9 Cross-section through the upper part of a field emission gun: C, cathode; W, Wehnelt electrode; A, anode. Also shown are the elements used for potential calculation by the CSM: four thin circular apertures, charged rings marked by dots, a homogeneously charged cylinder of radius R, an infinitely thin wire (located on the z-axis and consisting of two sections with constant but different charges per unit length) and a point charge Q_S. The control points at the surfaces are marked by crosses. *From Kasper (1979).*

the CSM associated with Hoch et al. (1978) can be thought of as a combination of CDM and methods from the earlier work of Regenstreif (1951), Glaser (1952), Lenz (1956b), Dommaschk (1965), Kanaya et al. (1966), and El-Kareh and El-Kareh (1970a), which were used to approximate a field with rotational symmetry produced by electrostatic lenses by means of superposition of some pure aperture fields.

The electron optics group in Tübingen (Kasper, 1979, 1981, 1982; Kasper & Lenz, 1980; Kasper & Scherle, 1982; Eupper, 1982) have continuously extended the CSM. First of all, Kasper (1979) added a coaxial cylinder, thin wires and a point charge to the coaxial rings and apertures and could thus simulate the field of an electron gun with a field emission tip. Fig. 9 shows a cross-section through this arrangement. Kasper and Scherle have generalized the CSM for Fourier series expansions like Eq. (40a). They introduced coaxial rings carrying a harmonic source distribution and applied this method to toroidal magnetic deflection systems with cylindrical ferrite yokes, in one case with asymmetric shielding. Eupper (1982) extended the CSM to three-dimensional problems. To do this, he used piecewise continuous-line segments with piecewise linear line charge densities inside the electrode. In this way, he was able to calculate the field in an electron gun with a cathode shaped like a hipped roof. An electron gun of this kind is capable of producing an electron line source.

Fig. 10 shows an example, calculated by the CSM, taken from the field of e-beam testing. The grounded and only roughly modeled DUT, parts of

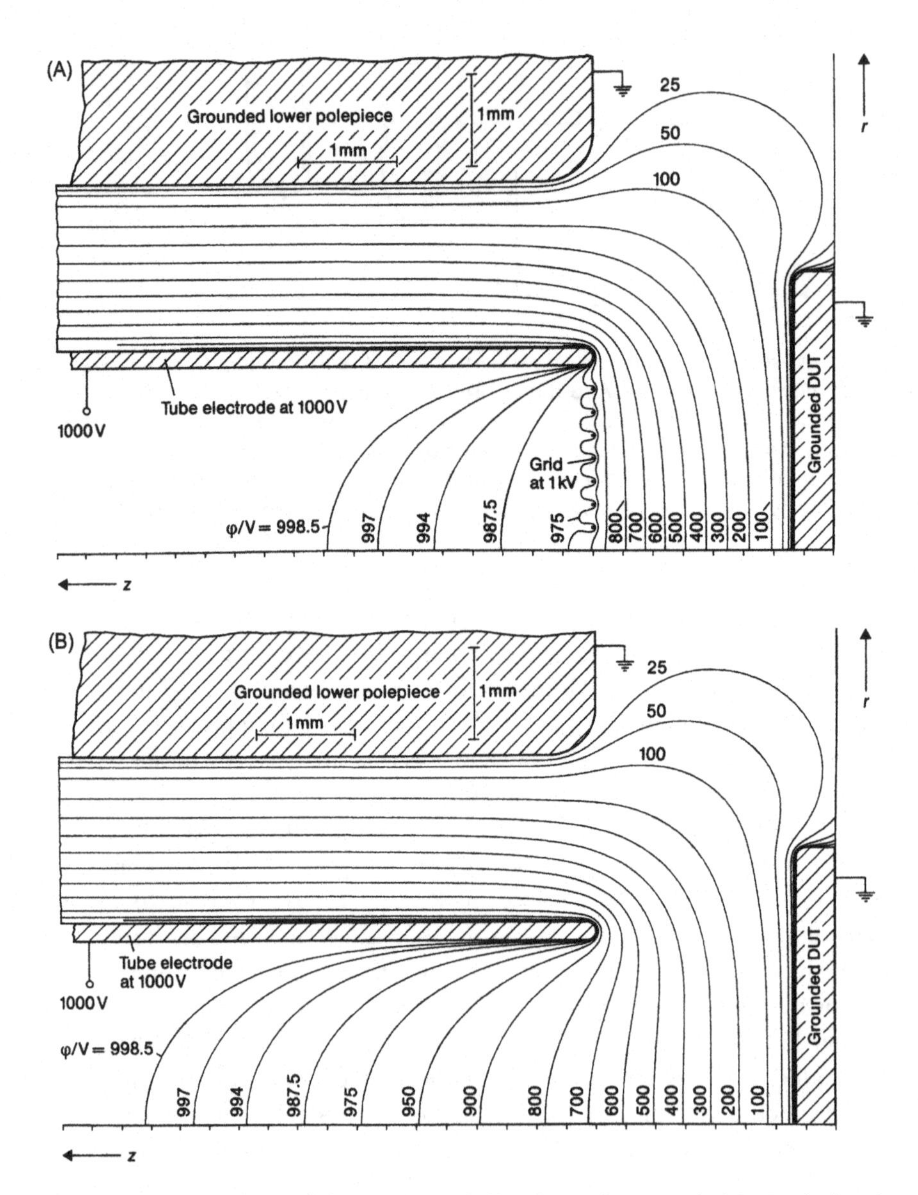

Figure 10 Potential lines of the extraction field in front of a grounded DUT calculated with the CSM program *SELECT* of T. Schwinn. In case (B) only a tubular electrode at 1 kV is used, in case (A) an additional mesh is arranged on the lower end of the tubular electrode. The grid comprises seven rings of 20 µm thickness. (C) A comparison of the axial potential $\varphi(0, 0, z) = \Phi(z)$ and the axial held strength $E_z(0, 0, z) = -\Phi'(z)$ for the two cases (A) and (B).

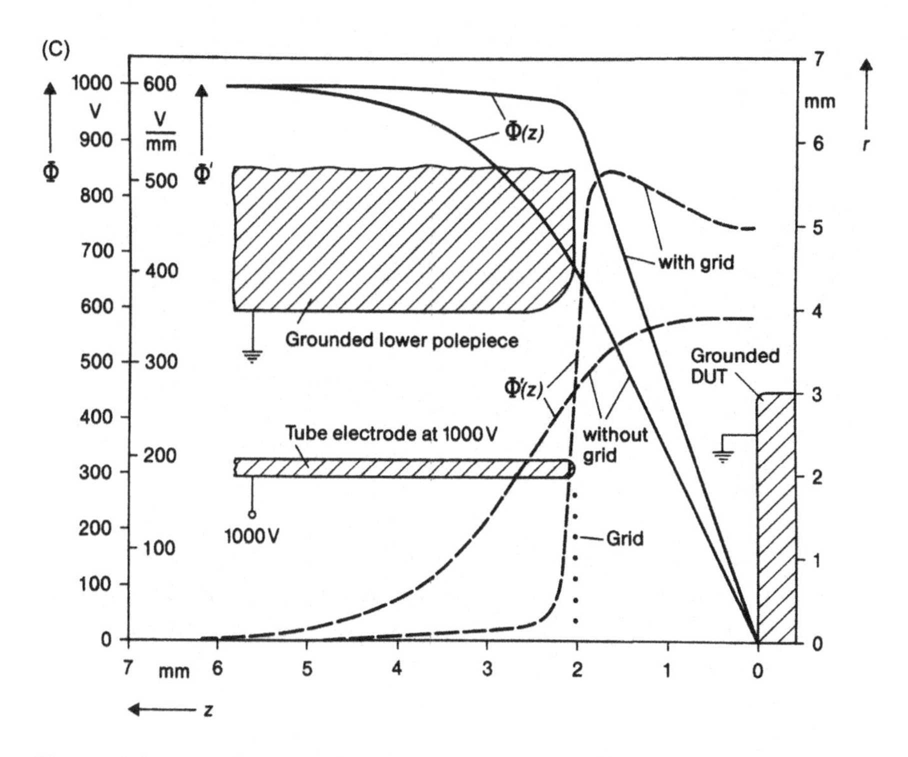

Figure 10 (*continued*).

the grounded lower magnetic pole piece and the lower part of the coaxial tubular electrode, which is here at a potential of 1 kV and is used to extract the SEs, are shown. Fig. 10A shows a mesh on the lower end of the tubular electrode; Fig. 10B does not. The point was to determine the degree to which the electrostatic extraction field at the DUT in case (A) exceeds that in case (B). To answer this question, exact modeling of the mesh is not so important. For this reason, the square grid was simulated by a free-hanging grid comprising seven rings of 20 μm thickness. For case (A), the extraction field at the DUT is greater by about 21%. It would certainly also have been possible to obtain this result by simulating a circular, conducting termination plate on the tubular electrode. The calculations for case (A), which were also performed using the FEM as a comparison, have also shown that the CSM gives extra-ordinarily continuous and smooth results in the grid region.

The equipotential lines in the tube shown in Fig. 10B act like a divergent lens on the primary electrons. Apart from the defocussing effect, which is not so crucial, additional deflection occurs of the PEs which have already been deflected further up by the scan coils. It was shown experimentally that the scan area for an arrangement shown by Fig. 10B was significantly greater than the diameter of the tube. The slightly weaker extraction field at the DUT produced by the arrangement without a grid is a small price to pay for this positive effect. If, on the other hand, the DUT is not a bonded and packaged chip but still part of a wafer whose bond pads have to be contacted by the wafer prober's "bed of nails", there are arguments in favor of the extraction grid. This is because if there were no grid the electrostatic field would mainly form between the edge of the tube electrode and the needles on the "bed of nails". This would strongly reduce the extraction field at the measurement point inside the chip.

Schönecker et al. (1986, 1990) have used line and sheet charges, slit apertures and arrays of line charges as substitute charge elements. This allowed them to simulate the local field (microfield) above the interconnections of ICs and calculate its influence on the trajectories of the SEs. Fig. 11 shows the potential lines and SE trajectories for an arrangement of three parallel interconnections which were simulated by Schönecker et al. (1986). Janzen (1990) took the dielectric under the interconnections into account when he simulated a similar arrangement.

Some Other Methods, Combined Methods, and a Comparison of Methods

We will now pass on to a short discussion of some publications on and techniques for calculating magnetic fields based on the IEM and the BEM. Using the BEM to handle magnetic problems is somewhat more problematic than dealing with electrostatic ones, especially when the permeability of the material under consideration is finite. It should also be mentioned that a variety of integral equations have to be solved if the reduced scalar potential, the total scalar potential, the vector potential or even the magnetization are considered; for references and discussions on these topics see Ströer (1987). We take a reduced scalar potential to mean that the magnetic field can be split up into a vortex part due to the coils and a source part due to the magnetic material. The vortex part can be calculated by appropriate integration using the Biot–Savart law. The source part is expressed in terms of the gradient of a (reduced) scalar potential and a Fredholm integral equation of the second kind is obtained for this scalar potential (Scherle, 1983). No matter what form the formulation and solution of the problem may take in a particular case, one feature common to all the BEMs is that only the surface and not the whole volume of the arrangement has to be dis-

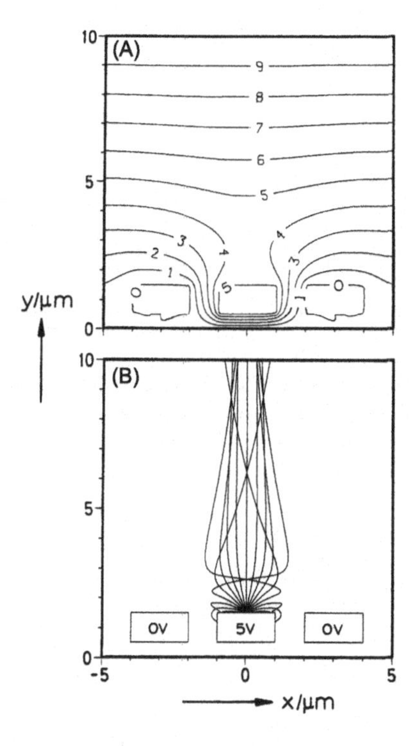

Figure 11 (A) Potential distribution and (B) SE trajectories for an arrangement of three (parallel and long) interconnection lines with a line width of 2 µm and a gap width of 1 µm. The extraction field is 1000 V mm^{-1}. The SEs start with an energy of 1.5 eV and the angle of emergence is increased in equal steps of 10°. *From Schönecker et al. (1986).*

cretized. This considerably reduces the rank of the system of equations to be solved. After solving this system of equations, the surface current densities are known and the field in whole space can be calculated by integration over the currents on material surfaces or in space.

The recent publications of Scherle (1983), Ströer (1987), Becker (1989, 1990), and Kasper and Ströer (1989, 1990) deserve mention as these authors have developed the use of the BEM for magnetic field calculations. We cannot discuss the numerical details (the collocation method, the Galerkin method, interpolation kernels, the method of least squares, etc.) as this would take us too far afield. The papers by Scherle (1983), Ströer (1987), Becker (1990), and Kasper and Ströer (1990) also deal with finite permeabilities μ, but are largely devoted to regions of different constant permeability. Only Kasper and Ströer (1990) discuss saturation effects ($\mu_r = \mu_r(H) \ggg 1$). These authors have performed calculations for round

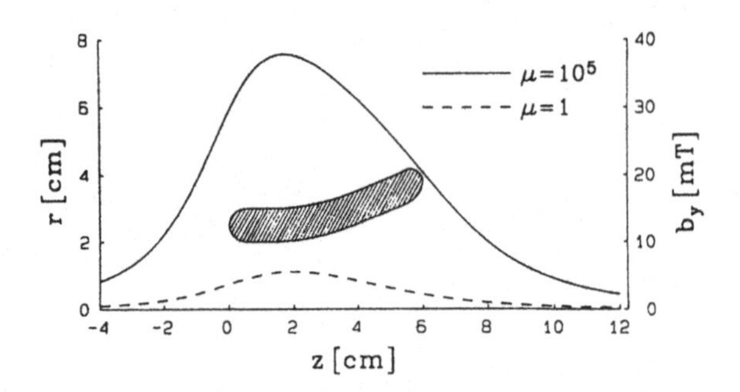

Figure 12 Axial magnetic deflection field distribution ($b_y = -\Psi_{1s}$) and half of the cross- section through the yoke of a toroidal coil system. The current-conducting loops, not shown here, are at a distance of 1 mm from the surface. For winding distribution and excitation current, see text. *From Kasper and Ströer (1989).*

lenses with large gap, single pole piece lenses, quadrupole lenses, bending magnets and deflection systems with ferromagnetic yokes and toroidal coils or saddle coils. Fig. 12 shows the axial distribution of the deflection field strength in a toroidal coil system together with half of a cross-section through the yoke. The current loops are not shown; they are 1 mm from the yoke surface. Moreover, a piecewise-constant winding distribution was assumed, the values being 1250 At between $\vartheta = 120$ and $240°$ and -1250 At between -60 and $+60°$. The reader who is interested in further magnetic field calculations by means of the BEM should consult the publications by Lucas (1976), Simkin and Trowbridge (1976), Iselin (1981), and Kuroda (1983), and the review articles by Kasper (1982, 1984a, 1987).

In the paper by Kasper and Ströer (1990), finite permeability is simulated by combining the FEM and the BEM. First, the FEM is used to calculate the vector potential over space, but only the values for the surfaces of the yokes and the coils are retained. For the remaining vacuum domain, the field calculation is thus reduced to a simple Dirichlet problem which is solved by the BEM.

Another reason for using combined or hybrid methods is the avoidance of artificial boundaries which must always be introduced when the FDM and the FEM are used. This is why Lencová and Lenc (1984) combined the FEM and the BEM. Since the BEM/CDM has disadvantages when space charge is present (Poisson's equation) and the FDM has difficulties with curved boundaries, Kasper (1984b) has proposed combining both methods in a way that would avoid their disadvantages. The results of Killes's (1985) test on the combination were successful.

If the FDM or the FEM are used, the potential is only obtained at the lattice points. For ray tracing, the potential and the field strength at an arbitrary point is usually required and this need not necessarily be a lattice point. This makes additional interpolation and differentiation necessary. References on this can be found in Kasper and Lenz (1980), Kasper (1982, 1987), Hawkes and Kasper (1989a), and Khursheed and Dinnis (1989).

In the publications by Janse (1971), Kasper (1982), Franzen and Munro (1987), Munro (1988), and Kasper and Ströer (1990), hints are given on the numerical calculation of fields arising in systems with rotational symmetry, whose electrodes or pole pieces may not have perfect rotational symmetry or are slightly misaligned. The use of the Fourier series method in combination with any numerical method appears to be the most reasonable procedure under these circumstances. Only multipole components up to the hexapole ($\nu = 3$ in Eqs. (5) or (40)) are of practical interest.

All numerical field calculation methods have their special strengths and characteristic weaknesses. No method on its own can be entirely satisfactory, nor can any single method be used for all applications. Which method is the most suitable must be decided from case to case and this depends very strongly on the details of the field calculation problem under consideration. Comparisons of the FDM and the FEM are found in Kasper and Lenz (1980) and Kasper (1982). The CDM, FDM, FEM and two analytical methods are compared by Mulvey and Wallington (1973), while Hawkes and Kasper (1989a) turn their attention to the BEM (CDM, CSM), FDM, and FEM. Mulvey and Wallington (1973) have produced a very clear and interesting table and the author would urge the reader to take the time to look it up because it is still valid today (NB $\mu \neq \infty$ can be accommodated by today's BEM). The author personally feels that the assessment factors of working memory and execution time play only a minor role in the majority of practical field calculation problems, even when PCs are used. This is because nowadays a PC with a 20 MHz processor, an additional numeric processor, and a 4 Mbyte working memory (which can in many cases be fully exploited only with an operating system extender) is a standard configuration.

Problems with computer time occur with true three-dimensional problems when, for example, many mesh points have to be used on the three-dimensional FDM solvers by Rouse and Munro (1989) or when any numerical field calculation has to be frequently repeated because of the lack of parametrization during optimization. Similarly, long computing times may occur when ray tracing is performed using a field calculated by means of the

CDM/CSM when a large number of subelectrodes or equivalent charges respectively have to be used and their contributions to each point on the trajectory have to be repeatedly summed. Often, when computing times for the CDM/CSM are being compared, it is clear that only the time to determine the equivalent charges is used. The code is simplest and shortest for the FDM.

The reasons for the great advances made in recent years are, first, the continual improvement and extension of the BEM and, secondly, the advantageous combination of the BEM with the (spatially) discrete FDM and FEM (see above). The author considers these combined methods to be of particular importance, but unfortunately, there are to his knowledge no easily obtainable programs that implement these combined methods. If the author were forced to say which method deserved first place, he would nominate the BEM (CDM/CSM) because:

- there is no need to introduce artificial boundaries, i.e. the fringing field is well described;
- this method can accommodate electrodes and boundaries with features that vary greatly in size;
- only the boundaries need to be discretized (modest working memory);
- because of the analytical form of the solution, it is automatically smooth and continuous, thus making numerical interpolation and differentiation superfluous.

To be honest, this method also has disadvantages. The BEM is inadequate if the boundary surfaces are "large" in total surface area. This means that the rank of the system of equations to be solved is a big number and the time taken to calculate potentials and fields is also very long. Eupper (1982) states that the disadvantages of the BEM/CDM are the difficulty in covering an arbitrary surface with plane polygons, setting up equations for the charge densities—which involves complicated case-by-case analysis—and the fact that there are no analytical expressions for the potential for any triangles or squares even when dealing with homogeneously charged area elements. Because of the advantages offered by the adaptability to complicated boundaries and even more so because saturation effects can be considered, the FEM is a firmly established technique. However, the problems encountered when the FEM is used to calculate field strength (Kasper & Lenz, 1980) must be borne in mind. The FDM is particularly useful when the lines of a square lattice coincide with the boundary. The comparison of methods given by Hawkes and Kasper (1989a), in the author's opinion, comes down a bit heavily on FDM, as he has been pleased with the performance of the three-dimensional FDM programs developed by Rouse and Munro (1989).

3. ELECTRON TRAJECTORIES

After the electric and/or magnetic field has been determined, the next step in an electron optical calculation is to find the electron trajectories. Starting with the Lorentz equation of motion, or the exact trajectory equation, a distinction can be made between two main methods: the perturbation method and the direct (usually numerical) ray-tracing method. When the perturbation method is used, the exact trajectory equation is linearized for small distances and trajectory inclinations to the optical axis ($|w|, |w'| \ll 1$). The analytic or numerical solution of the Gaussian trajectory equation obtained in this way (first-order trajectory equation) gives the principal trajectories, the cardinal elements and the fundamental trajectories. The next step of the perturbation theory is applied to obtain deviations from the fundamental trajectories in terms of aberrations. This classical method is, for instance, not suitable for calculating trajectories in electron guns or determining SE trajectories in spectrometers and detectors.

The starting point for ray tracing is the Lorentz equation of motion or a trajectory equation which has been derived from it. For certain initial values, the three space coordinates of the trajectory are calculated as a function of time, or some other trajectory parameter, e.g. arc length, using numerical methods. It is also possible to choose the coordinate of the optical axis as an independent variable and so determine $x(z)$ and $y(z)$. Using highly accurate numerical trajectory integration, and further measures that we will discuss later, it is even possible to determine the aberration coefficients from the intersects x_B and y_B of a set of electron trajectories in an image plane $z = z_B$. After a section dealing with a number of useful fundamental laws, we discuss the classical electron optical perturbation method with first-order optics and aberrations, go on to deal with ray tracing and, finally, discuss the optimization of electron optical components and systems. As we are only concerned with low-voltage optics at this juncture, we will restrict ourselves to non-relativistic approximations.

3.1 Equation of Motion and Some Fundamental Laws

The well-known non-relativistic approximation to the Lorentz equation of motion is given by

$$m\frac{\mathrm{d}\mathbf{v}}{\mathrm{d}t} = -e(\mathbf{E} + \mathbf{v} \times \mathbf{B}) \tag{53}$$

where t is time, $m = m_0$ ($= 9.1 \times 10^{-31}$ kg) is the rest mass of the electron, $-e$ the charge of an electron, and $\mathbf{v}$ its velocity. (We call $e = 1.6 \times 10^{-19}$ A s the elementary charge.) Another well-known result is the law of conservation of energy in the form

$$\frac{m}{2}v^2 - e\varphi = C_\mathrm{E} = \text{constant} \tag{54}$$

which is derived from Eq. (53). If φ is again the acceleration potential referred to the cathode, the constant C_E is the initial kinetic energy of the electron at the cathode. $C_\mathrm{E} = 0$ means that the energy spread of the electrons is neglected. When Eq. (54) is derived from Eq. (53), it is assumed that the fields do not vary with time. This, however, is not a valid assumption for, say, the beam-blanking system of an EBT. In this case, the equation

$$\frac{\mathrm{d}}{\mathrm{d}t}\left(\frac{m}{2}v^2 - e\varphi\right) = -e\frac{\partial\varphi}{\partial t} \tag{55}$$

as derived, for example, by Sommerfeld (1964), or Lischke, Plies, and Schmitt (1983), applies.

Lischke et al. (1983) have estimated the energy which is added to or taken from a primary electron which is inside a beam-blanking capacitor at the moment when the switching edge of the electric pulse is applied on the plates. This effect does not lead to an effective widening of the energy spread of the PEs if the pulse width is large in comparison with the rise time and fall time of the pulse.

Another important conserved quantity is the z-component of the canonical angular momentum $L_{\mathrm{can},z}$ in fields with rotational symmetry. If cylindrical coordinates r, ϑ, z are used then

$$L_{\mathrm{can},z} = mr^2\dot{\vartheta} - erA_\vartheta = C_\mathrm{L} = \text{constant} \tag{56}$$

where $\dot{\vartheta}$ has been written for the derivative of ϑ with respect to time $\mathrm{d}\vartheta/\mathrm{d}t$. The derivation of Eq. (56) may be found, for example, in Szilagyi (1988). The constant C_L is the initial value of the z-component of the canonical or generalized angular momentum and $C_L = 0$ for electrons that cross the z-axis during their motion. Using Stokes's theorem for the vector potential **A** (which has a ϑ-component only in fields with rotational symmetry), Eq. (56) takes the form of Busch's theorem:

$$mr^2\dot{\vartheta} - \frac{e}{2\pi}F = C_L \tag{57a}$$

where

$$F = 2\pi \int_0^r B_z r\mathrm{d}r \approx \pi r^2 B(z) \tag{57b}$$

is the magnetic flux which passes through a circle of radius r centered on the z-axis. For electrons with $C_L = 0$, the angular velocity $\dot{\vartheta}$ is the local Larmor frequency $\dot{\vartheta} = eB/(2m)$. If the field is purely electrostatic, the magnetic flux, F, is equal to zero and the kinetic angular momentum $mr^2\dot{\vartheta}$ is itself a constant.

If we examine the special case of a magnetic field which is varying slowly in space and time, i.e. adiabatically, with

$$\left|\frac{\partial \mathbf{B}}{\partial z}\right| \ll \frac{e\mathbf{B}^2}{mv} \quad \text{paraxial: } |B'| \ll |B|/r \tag{58a}$$

$$\partial|\mathbf{B}|/\partial t \ll e|\mathbf{B}|^2/m \tag{58b}$$

the kinetic angular momentum $mr^2\dot{\vartheta}$ is a constant to the first order of approximations (see Morrison, 1961; Cap, 1975; Jackson, 1975; or Artsimowitsch and Sagdejew, 1983). Then, according to Eq. (57a), the flux enclosed by the particle during its motion is also constant, i.e. the particle is spiraling along the surface of a flux tube of constant flux F. If α is the angle between the velocity vector and the z-axis, the following relationship holds for adiabatic particle motion:

$$\frac{\sin\alpha}{\sin\alpha_0} = \sqrt{\frac{B}{B_0}} = \frac{r_0}{r} \tag{59}$$

where the index 0 indicates the initial values (Morrison, 1961; Klemperer, 1972; Artsimowitsch & Sagdejew, 1983).

Using the conservation of the enclosed magnetic flux, the kinetic angular momentum or the magnetic moment of particle motion ($|\mathbf{m}| = er^2\dot{\vartheta}/2$), it is possible to explain the magnetic bottle and the magnetic mirror which are well-known phenomena encountered in plasma physics and geomagnetism (Morrison, 1961; Klemperer, 1972; Jackson, 1975; Artsimowitsch & Sagdejew, 1983), the photoelectron spectrometer (Beamson, Porter, & Turner, 1980), the magnetic field parallelizer (Kruit & Read, 1983) and the magnetic collimation of SEs for e-beam testing (Garth, Nixon, & Spicer, 1986; Kruit & Dubbeldam, 1987; Muray & Richardson, 1987). In the case of e-beam testing, the IC is located at a point where there is a strong magnetic field B_0 which decreases adiabatically along the direction of motion of the SEs so that, according to Eq. (59), the angle of its trajectory with respect to the z-axis continually decreases and its radius continually increases.

A very important fundamental law relating to charged-particle optics is Liouville's theorem which states that the volume of the phase space (formed from position coordinate space and momentum space) is constant during the motion of the particle. The Helmholtz theorem, as a paraxial approximation, and the conservation of normalized brightness follow from Liouville's theorem. We will discuss Helmholtz's theorem in the next section. We will take the normalized brightness to be the ratio of β/Φ, where the brightness $\beta = j/\Omega$, of an electron beam is defined as the current density j per unit solid angle Ω. Following our previous convention, Φ is the axial potential referred to the cathode, i.e. the beam voltage. According to Langmuir (1937), the maximum obtainable brightness is given by the following relationship:

$$\beta \approx \frac{j_c}{\pi} \frac{e\Phi}{kT_c} \tag{60}$$

where k is Boltzmann's constant, and j_c, T_c are the current density and absolute temperature at the cathode, respectively. In low-voltage electron optics in particular, this value is not reached. The reasons for this are saturation effects in the gun, aberrations and, especially, beam broadening due to electron–electron interaction.

All conservation laws such as Eq. (54) or (56) can be used as useful extra checks on the accuracy of numerical trajectory computations (e.g. see Plies & Schweizer, 1987). Sometimes it is worthwhile remembering that the particle model is not the only way of interpreting electrons, as the wave model is equally valid. The electron trajectories are the normals to the wave surface if $\mathbf{A} = 0$. By using the electron optical refractive index

$$n = \frac{1}{mc}\left(mv - \frac{e\mathbf{A}\cdot\mathbf{v}}{v}\right) \tag{61}$$

and Fermat's principle, it is even possible to obtain the extra information that certain aberration coefficients are real and/or satisfy certain relations. The eikonal method, as this procedure is called, was first applied to electron optics by Glaser (1933). The relationships between the aberration coefficients which are obtained from this method can be used to check the accuracy of the analytic or numerical results which have been obtained by using the trajectory method. We will return to this topic in the next section but one.

Sometimes, it can be helpful to use the well-known electron optical similarity theorems when dealing with certain electron optical problems. We will simply refer to Glaser (1952, 1956), Szilagyi (1988), or Hawkes and Kasper (1989a). The following formula, derived, for example, by Glaser (1956),

$$\varkappa = \frac{1}{\rho} = -\frac{E_{\mathrm{n}}}{2\varphi} + \sqrt{\frac{e}{2m\varphi}}B_{\mathrm{b}} = -\frac{eE_{\mathrm{n}}}{mv^2} + \frac{eB_{\mathrm{b}}}{mv} \tag{62}$$

is very useful when considering the curvature $\varkappa$ of electron trajectories. The symbols ρ, E_{n}, and B_{b} represent the radius of curvature, the electric field strength along the principal normal, and the magnetic flux density along the binormal of the electron trajectory, respectively.

The Lorentz equation (53) is not always convenient. In many practical applications the trajectory is not required in the form $\mathbf{r} = \mathbf{r}(t)$, i.e. position vector as a function of time, a function of the space coordinates, e.g. $x = x(z)$ and $y = y(z)$, being preferable instead. Therefore, the time parameter t is often eliminated and a more convenient parameter used, e.g. the arc length s or the axial coordinate z. Glaser (1956) and Hawkes and Kasper (1989a) state very general trajectory equations with s as the parameter. A trajectory equation with z as the parameter can be obtained from Eq. (53) using

$$\frac{\mathrm{d}}{\mathrm{d}t} = \frac{\mathrm{d}s}{\mathrm{d}t}\frac{\mathrm{d}z}{\mathrm{d}s}\frac{\mathrm{d}}{\mathrm{d}z} = \frac{v}{\sqrt{1 + x'^2 + y'^2}}\frac{\mathrm{d}}{\mathrm{d}z} = \sqrt{\frac{2e\varphi}{m}}\frac{1}{\sqrt{1 + x'^2 + y'^2}}\frac{\mathrm{d}}{\mathrm{d}z}$$

Employing the complex number notation introduced in Eqs. (2) and (4), Eq. (53) can be written as

$$\frac{\mathrm{d}}{\mathrm{d}z}\left(\sqrt{\frac{\varphi}{1 + w'\overline{w}'}}w'\right) = -\frac{1}{2}\sqrt{\frac{1 + w'\overline{w}'}{\varphi}}E_w + \mathrm{i}\sqrt{\frac{e}{2m}}(w'B_z - B_w) \tag{63}$$

This trajectory equation for $w = w(z) = x(z) + \mathrm{i}y(z)$ is well suited for developing a systematic theory of focusing, first-order deflection and aberrations and was used by Chu and Munro (1982a) and in its relativistically corrected form by Plies (1982a). When the trajectory equation (63) is derived from the equation of motion (53), the following equation is also obtained:

$$\frac{\mathrm{d}}{\mathrm{d}z}\left(\sqrt{\frac{\varphi}{1+w'\overline{w}'}}\right)=-\frac{1}{2}\sqrt{\frac{1+w'\overline{w}'}{\varphi}}E_z+\sqrt{\frac{e}{2m}}\mathrm{Re}\left(\mathrm{i}B_w\overline{w}'\right) \tag{64}$$

If Eqs. (63) and (64) are combined, we get a further trajectory equation:

$$\begin{aligned}w''=&\frac{1+w'\overline{w}'}{2\varphi}(-E_w+w'E_z)\\&+\mathrm{i}\sqrt{\frac{e}{2m\varphi}}\sqrt{1+w'\overline{w}'}\left(-B_w+w'B_z-\frac{1}{2}w'\overline{w}'B_w+\frac{1}{2}w'^2\overline{B}_w\right)\end{aligned} \tag{65}$$

Unlike Eq. (63), this differential trajectory equation is well suited for numerical integration. It has already been published in this form, e.g. by Kasper (1987), and also corresponds to the two equations for x'' and y'' given by Hawkes and Kasper (1989a, Eqs. (3.22)), if the abbreviations used there are taken into account. It should be noted that only a straight optical axis was assumed for the trajectory equations (63) and (65) and that no further assumptions were made, e.g. about the symmetry of the field.

When the electron trajectories are to be found using numerical methods, the question arises as to which form of the equation of motion, or the trajectory equation, is the most suitable. There is no universal answer to this question because the choice of a suitable parameter depends on whether a numerical trajectory computation has to be performed, for example, in an electron gun or an electron mirror, or a magnetic lens, which is, in contrast, well behaved. The selected parameter should remain within a reasonable range throughout the whole trajectory integration, and the step width for numerical integration should not vary too much either, so as to maintain the required accuracy. Kasper (1985) suggested a time-like parameter σ,

$$\mathrm{d}\sigma=\sqrt{\frac{2e}{m}\varphi_{\mathrm{R}}}\,\mathrm{d}t \tag{66}$$

which is usually satisfactory, φ_{R} represents the constant acceleration potential at a suitable reference point. Using this parameter, σ, which has the dimension of length, Kasper transformed the Lorentz equation and obtained

$$\frac{\mathrm{d}^2\mathbf{r}}{\mathrm{d}\sigma^2}=\frac{1}{2}\nabla\left(\frac{\varphi}{\varphi_{\mathrm{R}}}\right)+\sqrt{\frac{e}{2m\varphi_{\mathrm{R}}}}\mathbf{B}\times\frac{\mathrm{d}\mathbf{r}}{\mathrm{d}\sigma} \tag{67}$$

We will return to this equation in the section on ray tracing. It should be mentioned that in Eq. (66) Kasper used the relativistic proper-time element $\sqrt{(1 - v^2/c^2)}\mathrm{d}t$ instead of the time interval dt. However, this is not necessary for low-voltage electron optics.

3.2 First-Order or Paraxial Optics

Before discussing first-order optics *per se*, let us make certain restrictions on the electrostatic and magnetic fields with which we will be dealing.

This and the following section largely follow the methods developed by Plies (1982a, 1982b) and by Plies and Elstner (1989a, 1989b) as far as notation and content are concerned; these publications, however, are based on a full relativistic treatment, which is not required here.

The SEM, and its variations for e-beam lithography, e-beam testing and electron optical inspection, use deflection elements for beam scanning as well as lenses of rotational symmetry. Scanning electron systems that contain quadrupole lenses for focusing beams, or to correct aberrations, are rare today and will not be discussed here; however, see Smith and Munro (1986) or Okayama (1990). For pure quadrupole focusing systems, see the two monographs by Hawkes (1966, 1970). We will assume two planes of symmetry for beam deflectors, i.e. that the potential is antisymmetrical along the deflection field and symmetrical perpendicular to it (see Fig. 2). Under these assumptions, the deflection fields have no even multipole components and, in particular, no quadrupole components, i.e. $\varphi_{2\nu} = \Phi_{2\nu} = \psi_{2\nu} = \Psi_{2\nu} = 0$ for $\nu \geqslant 1$. Moreover, we consider only deflection elements with spatially superposed x- and y-deflection, whose deflection electrodes and coils have the same form for both types (see Fig. 2B, C). The potential of deflection elements of this kind is invariant when the coordinate system and the deflection voltages or currents are both rotated through 90°.

This invariance can be expressed as

$$\left.\begin{aligned} \varphi(\mathrm{i}w, -\mathrm{i}\overline{w}, z, \mathrm{i}U, -\mathrm{i}\overline{U}) &= \varphi(w, \overline{w}, z, U, \overline{U}) \\ \psi(\mathrm{i}w, -\mathrm{i}\overline{w}, z, \mathrm{i}I, -\mathrm{i}\overline{I}) &= \psi(w, \overline{w}, z, I, \overline{I}) \end{aligned}\right\} \tag{68a}$$

if the complex deflection voltages and currents

$$U = U_x + \mathrm{i}U_y \quad \text{and} \quad I = I_x + \mathrm{i}I_y \tag{68b}$$

are used. If the deflection voltages and currents are real, the index indicates the deflection field direction which only coincides with the direction of the beam deflection in the electric case.[1] By separating U and I, the dipole and hexapole functions can be written using new normalized axial functions:

$$\Phi_1(z) = -P_1(z)U \qquad \Phi_3(z) = -\frac{1}{3}P_3(z)\overline{U} \tag{69a}$$

$$\Psi_1(z) = -Q_1(z)I \qquad \Psi_3(z) = -\frac{1}{3}Q_3(z)\overline{I} \tag{69b}$$

The axial functions and the deflection voltages and currents which vary with time are separated by these two equations.

If multistage beam deflections with principal deflection planes which are rotated with respect to each other are considered, the normalized axial functions P_1, P_3, Q_1, and Q_3 are complex. In this case, U and I can be interpreted as, say, the deflection voltage or deflection current of the first deflection element for which the normalized axial function can be assumed to be real. The total normalized distributions then apply for a specific adjustment of the different deflection stages (ratio of the absolute values of the individual deflection voltages or currents) and for a specific azimuthal orientation of the individual deflectors with respect to each other.

If all higher-order multipole components, except for the field with rotational symmetry and the dipole field vanish, i.e. $\Phi_\nu \equiv \Psi_\nu \equiv 0$ for $\nu > 1$, then we refer to a field with generalized uniformity (Ohiwa, Goto, & Ono, 1971) or a rotation invariant potential (Goto & Soma, 1977; Soma, 1977). Then, in addition to Eq. (68a), the potentials are invariant with respect to a rotation of the coordinate system and a rotation of the deflection voltages or currents through any angle. If only $\Phi_2 \equiv \Phi_3 \equiv \Psi_2 \equiv \Psi_3 \equiv 0$, then rotation invariance holds up to and including the third order, which is sufficient for most components and systems used in practice.

Now, we come to the real theme of this section on first-order optics. We take the exact trajectory equation (63) as our starting point and, following Scherzer (1937), use a trajectory construction based on perturbation methods. In other words, we will substitute the multipole expansion (5), from Section 2.1 in Eq. (63), take Eqs. (4), (69a), and (69b) into account, expand the square roots and arrange the quantities w, $\overline{w}$, w', $\overline{w}'$, w'', $\overline{w}''$, U,

[1] This definition is different from that given by Chu and Munro (1982a). They define I_x as the current which deflects the beam along the x-axis of an individual deflector. The definition given here, however, follows Munro (1974) so that I_x produces a deflection field in the x-direction. The normalized Cartesian axial function chosen by Munro (1974) implies the relationships $Q_1 = D$, $-Q_3 = E + D''/8$.

$\overline{U}$, I, and $\overline{I}$, which are assumed to be small, according to their powers. This leads to the trajectory equation

$$w'' + \left(\frac{\Phi'}{2\Phi} - \mathrm{i}\sqrt{\frac{e}{2m\Phi}}B\right)w' + \left(\frac{\Phi''}{4\Phi} - \frac{\mathrm{i}}{2}\sqrt{\frac{e}{2m\Phi}}B'\right)w = r_1 + \Sigma \tag{70a}$$

with

$$r_1 = -\frac{P_1}{2\Phi}U - \mathrm{i}\sqrt{\frac{e}{2m\Phi}}Q_1 I \tag{70b}$$

and

$$\Sigma = s_3 + s_{2c} + s_{1J} + s_{2J} + s_{2Jc} + s_{3\mathbf{K}} \tag{70c}$$

The homogeneous part of this differential trajectory equation ($r_1 = \Sigma = 0$) describes focusing by round electrostatic lenses ($\Phi' \neq 0$) and round magnetic lenses ($B \neq 0$). The first-order beam deflection is determined by means of the linear inhomogeneous differential equation ($\Sigma = 0$) where Σ is a sum of non-linear perturbation terms which determine the image aberrations. These perturbation terms have already been derived in an explicit form by Plies (1982a). They will not be restated here, but their physical interpretation will be discussed in Section 3.3. It should be stressed that the derivation of Eqs. (70a)–(70c) with respect to the field only takes the assumed two planes of symmetry of the deflection elements into account. We have assumed nothing about the spatial arrangement of the lenses and deflection fields.

First-Order Focusing

First, we introduce two (known) coordinate transformations to simplify the homogeneous, linear, trajectory equation. Nowadays, this makes it possible to use simpler and more accurate methods of numerically determining the fundamental focusing trajectories whereas previously it made it easier to find any analytic solutions that might exist. By means of the first transformation

$$\left.\begin{aligned} &r(z) = w(z)\mathrm{e}^{-\mathrm{i}\chi(z)} \\ &\text{where} \\ &\chi = \sqrt{\frac{e}{8m}}\int_{z_G}^{z}\frac{B}{\sqrt{\Phi}}\,\mathrm{d}z \end{aligned}\right\} \tag{71}$$

the complex w-trajectory is transformed into a real trajectory r. This procedure effects a transformation into a coordinate system rotated about the

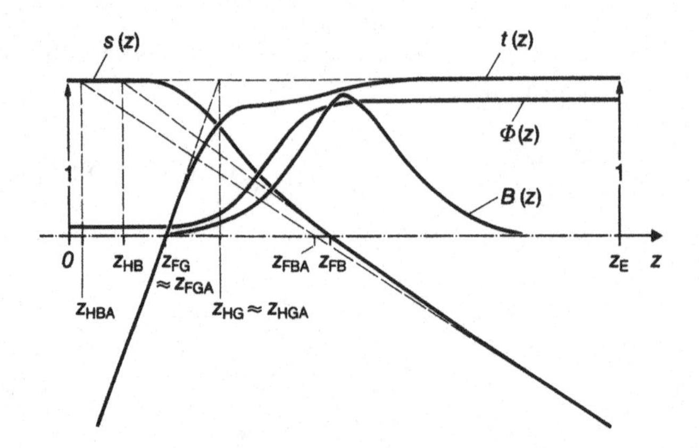

Figure 13 The principal trajectories of the focusing $s(z)$ and $t(z)$, from which the positions of the foci and principal planes are determined. z_{HB}, z_{FB}, z_{HBA}, and z_{FBA} are the image side positions of the real principal plane, the real focus, the asymptotic principal plane and the asymptotic focus, respectively. z_{HG}, z_{FG}, z_{HGA}, and z_{FGA} are the same cardinal elements on the object side. In this special example, the real and asymptotic cardinal elements coincide on the object side. *From Plies and Elstner (1989a).*

optical axis by the Larmor angle χ.[2] The lower limit of the integral for χ is chosen as the coordinate of the object plane z_G, and the result $\chi(z_B)$ in the image plane z_B is therefore the image rotation.

Since we also wish to include round electrostatic lenses, it is useful to also subject the trajectory $r(z)$ to the Picht transformation (Picht, 1957) in the form

$$R(z) = r(z)\left[\Phi(z)/\Phi(z_T)\right]^{1/4} \tag{72}$$

As will be seen further on, it is advantageous to make z_T in the equation the fixed z-coordinate at the beginning ($z = 0$) or the end ($z = z_E$) of the round lens fields (Fig. 13).

From Eq. (70a) with $r_1 = \Sigma = 0$ and by using the above-mentioned transformations, we obtain the linear homogeneous differential equations for the r-trajectory

$$r'' + \frac{1}{2}\frac{\Phi'}{\Phi}r' + \left(\frac{1}{4}\frac{\Phi''}{\Phi} + \frac{e}{8m\Phi}B^2\right)r = 0 \tag{73}$$

[2] $\dot{\chi} = \chi'\dot{z}(0, z) = eB(z)/(2m) = \omega_L =$ the local Larmor frequency, which is one-half of the local cyclotron frequency $\omega_c = eB/m$. The Larmor frequency is the angular velocity with respect to the optical axis, while the cyclotron frequency is the angular velocity with respect to the center of curvature of the trajectory. Unfortunately, some authors erroneously refer to eB/m as the Larmor frequency.

and for the R-trajectory,

$$R'' + \left(\frac{3}{16}\frac{\Phi'^2}{\Phi^2} + \frac{e}{8m\Phi}B^2\right)R = 0 \tag{74}$$

The last form does *not* contain the first derivative of the trajectory, even in the presence of round electrostatic lens fields ($\Phi' \neq 0$). Eq. (74) is better suited than Eq. (73) for numerical trajectory computation (see Section 3.4).

Principal and Fundamental Trajectories of the Focusing. In the w-, r-, and R-coordinate systems, the focusing is described by a linear homogeneous differential equation of the second order. Accordingly, there are always two linearly independent solutions in each case. Of special interest for the calculation of the cardinal elements in the r-coordinate system are the two principal trajectories $s(z)$ and $t(z)$ which satisfy the conditions

$$\left.\begin{array}{lll} s(0) = 1 & \text{and} & s'(0) = 0 \\ t(z_E) = 1 & \text{and} & t'(z_E) = 0 \end{array}\right\} \tag{75}$$

at the beginning ($z = 0$) and the end ($z = z_E$) of the round lens fields (electrostatic and/or magnetic) (see Fig. 13).

As can be seen from Fig. 13, the first principal ray $s(z)$ provides the image side position of the real and asymptotic focal and principal points. The second principal ray $t(z)$ does the same for the object side. In contrast to other fundamental rays, we call $s(z)$ and $t(z)$ principal rays because they provide the principal or cardinal elements. To calculate the image aberrations of a system with an aperture stop, we need two fundamental trajectories $r_\alpha(z)$ and $r_\gamma(z)$, which must satisfy the following conditions in the object plane z_G and in the aperture stop plane z_A (Fig. 14):

$$\left.\begin{array}{lll} r_\alpha(z_G) = 0 & \text{and} & r'_\alpha(z_G) = 1 \\ r_\gamma(z_G) = 1 & \text{and} & r_\gamma(z_A) = 0 \end{array}\right\} \tag{76}$$

If no aperture stop is present, $r_\gamma(z)$ is replaced by the trajectory $r^*_\gamma(z)$, where

$$r^*_\gamma(z_G) = 1 \quad \text{and} \quad r^{*\prime}_\gamma(z_G) = 0 \tag{77}$$

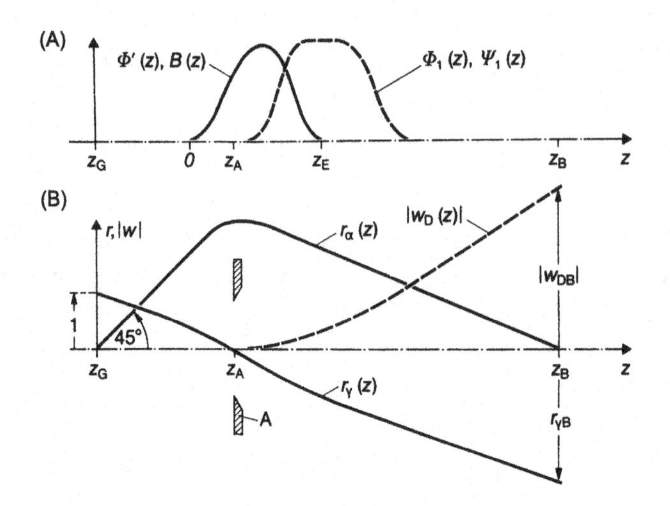

Figure 14 (B) The fundamental trajectories of the focusing $r_\alpha(z)$ and $r_\gamma(z)$ and the absolute value of the fundamental trajectory of the deflection $w_D(z)$, and (A) the associated axis functions $\Phi'(z)$, $B(z)$ of the rotationally symmetrical lens fields and $\Phi_1(z)$, $\Psi_1(z)$ of the deflection fields. A, aperture stop. Object position z_G and/or image position z_B may also be inside the field region. *From Plies and Elstner (1989a).*

If the trajectories s and t are already known, the trajectories r_α, r_γ, and r_γ^* are derived from them as a linear combination:

$$\left.\begin{aligned} r_\alpha &= \frac{s_G t - s t_G}{s_G t'_G - s'_G t_G} \\ r_\gamma &= \frac{s t_A - s_A t}{s_G t_A - s_A t_G} \\ r_\gamma^* &= \frac{s t'_G - s'_G t}{s_G t'_G - s'_G t_G} \end{aligned}\right\} \tag{78}$$

where $s_G = s(z_G)$, $t_A = t(z_A)$, etc.

In the R-system, we have to determine the two trajectories S, T with the initial conditions

$$\left.\begin{aligned} S(0) &= 1 \qquad & S'(0) &= \frac{\Phi'(0)}{4\Phi(0)} \\ T(z_E) &= 1 \qquad & T'(z_E) &= \frac{\Phi'(z_E)}{4\Phi(z_E)} \end{aligned}\right\} \tag{79}$$

In selecting these initial conditions, S is transformed to s and T to t when $z_T = 0$ and $z_T = z_E$ are selected for the transformation equation (72) in the first and second case, respectively. To be exact, therefore, two R-systems are available but they differ only by a constant factor.

The calculation then proceeds as follows. After numerical calculation of the paraxial trajectories S and T, the principal rays s and t are determined by the transformation equation (72), the fundamental trajectories r_α, r_γ and r_γ^* by the transformation equations (78) and, finally, the paraxial trajectories w_α and w_γ or w_γ^* in the laboratory system by the transformation (71). For completeness, let us also specify the conditions for the latter trajectories:

$$\left.\begin{array}{ll} w_\alpha(z_G) = 0 & w'_\alpha(z_G) = 1 \\ w_\gamma(z_G) = 1 & w_\gamma(z_A) = 0 \\ w_\gamma^*(z_G) = 1 & w_\gamma^{*\prime}(z_G) = \sqrt{\dfrac{e}{8m\Phi(z_G)}}B(z_G) \end{array}\right\} \quad (80)$$

For transformation (71), the Larmor angle χ must previously be determined by some numerical integration algorithm, e.g. by applying a combination of Simpson's rule and Newton's 3/8 rule.

A general trajectory of first-order focusing can now be written as

$$w_R(z) = \alpha w_\alpha(z) + \gamma w_\gamma(z) \quad (81)$$

The index R indicates the rotational symmetry of the imaging fields. The complex trajectory parameters $\alpha = w_{RA}/w_{\alpha A}$ and $\gamma = w_{RG}$ are obtained from the points of intersection of the trajectory w_R with the aperture or the object plane. In this case, α is the object side aperture and γ characterizes the lateral dimension of the object.

The above way of calculating the fundamental trajectories from the principal trajectories s and t in terms of linear combinations makes sense as s and t are required in any case to calculate the cardinal elements. It is feasible for the fundamental trajectory combinations r_α, r_γ^* mentioned above (initial-value problem) and r_α, r_γ (mixed initial boundary value problem). If the α-trajectory is fixed using the boundary conditions $\tilde{r}_\alpha(z_G) = 0$ and $\tilde{r}_\alpha(z_A) = 1$, the pairs $\tilde{r}_\alpha$, r_γ^* (mixed boundary initial value problem) and $\tilde{r}_\alpha$, r_γ (boundary value problem) can also be calculated using linear combinations of the principal trajectories s and t. This cannot be done when the object plane $z = z_G$ and the image plane $z = z_B$ are fixed from the start and the boundary value problem with $r_\alpha(z_G) = r_\alpha(z_B) = 0$ must be solved. (The s- and t-trajectories must then fulfill the condition $s_G t_B - s_B t_G = 0$.) Therefore, it is better to apply standard "shooting methods", i.e. the initial-value problem $r_\alpha(z_G) = 0$ and $r'_\alpha(z_G) = 1$ is solved, and the lens excitation is readjusted from the value of $r_\alpha(z_B) \neq 0$ or the value of z_0 for $r_\alpha(z_0) = 0$ until $r_\alpha(z_B) = 0$. Hawkes and Kasper (1989a) deal with this topic and Kasper (1982) provides more information on magnetic lenses in this context. Lencová and Lenc (1989) describe a refined shooting method for a magnetic lens and give some critical comments on the method of Glaser (1952), who attempted to solve the boundary value problem which has been mentioned by evaluating the eigenvalues of the paraxial ray equation using the Rayleigh–Ritz variational method. It should be mentioned that, in the case of superposed electrostatic and magnetic lens fields, there are two, free, field strength parameters which can be varied.

For the fundamental trajectories w_α and w_γ (or w_γ^*), the Helmholtz–Lagrange relation in the form

$$\sqrt{\Phi}\left(\overline{w}_\gamma w'_\alpha - w'_\gamma \overline{w}_\alpha\right) = \sqrt{\Phi_G} \tag{82}$$

is applicable. After determining the trajectories w_α and w_γ numerically, this relation can be used as an additional accuracy check.

Cardinal Elements of the Focusing. The real and asymptotic foci and principal plane positions result from the *s*- and *t*-trajectories as shown schematically in Fig. 13. Since the trajectory ordinates are known only at a finite number of sample points, the zeros of the *s*- and *t*-trajectories—i.e. the real foci—must be determined more precisely by means of numerical interpolation. All the other ordinate values or derivatives at intermediate points must also be determined by interpolation, e.g. (cubic) Bessel interpolations, from the corresponding adjacent sample points. If multiple focusing occurs (not shown in Fig. 13), one of the real foci must be selected as the significant real focus. The real cardinal elements are relevant to objective operation, the asymptotic ones to projective operation (projective, intermediate and condenser lenses).

Useful formulae for the focus of purely magnetic lenses as a function of the pole piece bores, the pole piece gap and the lens excitation can be found in Durandeau and Fert (1957) and Dugas, Durandeau, and Fert (1961), and a collection of sets of curves can be found in Hawkes (1982). Picht (1957) gives simple approximate formulae for the foci of electrostatic aperture lenses, immersion lenses and Einzel lenses. Useful curves for the cardinal elements of purely electrostatic lenses can be found, for example, in the books by Harting and Read (1976) and Szilagyi (1988). The method of sectionizing the lens at least helps in obtaining an understanding of the action of an electrostatic lens (El-Kareh & El-Kareh, 1970a; Szilagyi, 1988).

In objective lens operation, the position of the image plane z_B is obtained as the zero point of the r_α trajectory. In the case of multiple imaging, the significant real image in the field can again be selected. In projector lens operation, the image plane z_B is obtained as the intersection of the optical axis with the tangent to the r_α trajectory at z_E. The lateral magnification V and the image rotation ϑ_B are conveniently calculated from the fundamental trajectory r_γ and its transform w_γ (or r_γ^* and w_γ^*) in line with

$$V = r_\gamma(z_B) = r_\gamma^*(z_B) \tag{83a}$$

and

$$\vartheta_B = \chi(z_B) = \begin{cases} \arg w_\gamma(z_B) - \pi = \arg w_\gamma^*(z_B) - \pi & \text{for } n = 0, 2, 4, \ldots \\ \arg w_\gamma(z_B) = \arg w_\gamma^*(z_B) & \text{for } n = 1, 3, 5, \ldots \end{cases} \tag{83b}$$

With this definition of the lateral magnification, V is not an absolute number, but has the sign $(-1)^{n+1}$ where n is the number of intermediate images. The angular magnification is equal to $r'_\alpha(z_B)$. In the case of projective operation, Eqs. (83a) and (83b) should be treated accordingly, i.e. $r_\gamma(z_B)$ results from the tangent at the field end, $\chi(z_B) := \chi(z_E)$ and $\arg w_\gamma(z_B) := \arg w_\gamma(z_E)$.

In practice, there are cases (e.g. the compound spectrometer objective lens of Frosien & Plies, 1987) for e-beam testing) of objective imaging at one side of the lens, but projector imaging at the other. The *EMAF* program of Plies and Elstner (1989a, 1989b) can also handle this mixed objective–projector operation, with the restriction that the objective side must be the object side. However, this is not a real restriction, as inverting the entry into the field distribution allows the object and the image sides to be interchanged in a purely formal sense.

First-Order Deflection

Plies (1982a) has compiled a list of over 30 selected publications on beam scanning with a straight optical axis covering the period 1938 to 1980. In most of these publications, it was assumed that the lens field and the deflection field did not interpenetrate. We will now go on to discuss the most general case of spatially superposed electric and magnetic focusing and deflection fields. Glaser (1952) has already discussed the general case but limited himself to the first-order properties.

Fundamental Trajectories of the Deflection. If two fundamental focusing trajectories, i.e. two solutions of the homogeneous part of the differential equation (70a) are known, the linear inhomogeneous differential equation with $r_1 \neq 0$ and $\Sigma = 0$ can be solved by the method of variation of parameters. If it is assumed that w_α and w_γ are known, the following is obtained for the deflection trajectory designated by w_D:

$$w_\mathrm{D} = \frac{w_\alpha}{\sqrt{\Phi_\mathrm{G}}} \int\limits_{z_\mathrm{G}}^{z} \sqrt{\Phi} r_1 \overline{w}_\gamma \mathrm{d}z - \frac{w_\gamma}{\sqrt{\Phi_\mathrm{G}}} \int\limits_{z_\mathrm{G}}^{z} \sqrt{\Phi} r_1 \overline{w}_\alpha \mathrm{d}z \tag{84}$$

Here the lower integration limit of the integrals in Eq. (84) was selected so that the deflection trajectory satisfies the initial conditions $w_\mathrm{D}(z_\mathrm{G}) = w'_\mathrm{D}(z_\mathrm{G}) = 0$ (see Fig. 14). If we insert r_1 from Eq. (70b) into Eq. (84), then the deflection trajectory w_D can be split up into an electric and a magnetic component:

$$w_\mathrm{D}(z) = U w_U(z) + I w_I(z) \tag{85a}$$

where

$$w_U = \frac{-1}{2\sqrt{\Phi_\mathrm{G}}} \left(w_\alpha \int\limits_{z_\mathrm{G}}^{z} \frac{P_1}{\sqrt{\Phi}} \overline{w}_\gamma \mathrm{d}z - w_\gamma \int\limits_{z_\mathrm{G}}^{z} \frac{P_1}{\sqrt{\Phi}} \overline{w}_\alpha \mathrm{d}z \right) \tag{85b}$$

$$w_I = -\mathrm{i}\sqrt{\frac{e}{2m\Phi_\mathrm{G}}} \left(w_\alpha \int\limits_{z_\mathrm{G}}^{z} Q_1 \overline{w}_\gamma \mathrm{d}z - w_\gamma \int\limits_{z_\mathrm{G}}^{z} Q_1 \overline{w}_\alpha \mathrm{d}z \right) \tag{85c}$$

The deflection trajectories w_U and w_I can now be determined from the known deflection field distributions $P_1(z)$ and $Q_1(z)$ and the previously calculated fundamental focusing trajectories w_α and w_γ simply by means of numerical integration.

The entire first-order trajectory is finally obtained by superposition of the focusing trajectory $w_\mathrm{R}(z)$ and the deflection trajectory $w_\mathrm{D}(z)$.

$$w_0(z) = w_\mathrm{R}(z) + w_\mathrm{D}(z) = \alpha w_\alpha(z) + \gamma w_\gamma(z) + U w_U(z) + I w_I(z) \tag{86}$$

Cardinal Elements of the Deflection. The cardinal elements of the deflection are the deflection sensitivity and the position of the principal deflection plane. It is useful to once more consider the electrical and magnetic deflection separately here. From the deflection trajectories $w_U(z)$ and $w_I(z)$, we obtain the electrical deflection sensitivity $|w_U(z_\mathrm{B})|$ and the magnetic deflection sensitivity $|w_I(z_\mathrm{B})|$. The angles $\arg w_{U\mathrm{B}}$ and $\arg w_{I\mathrm{B}}$ specify the electrical and magnetic deflection directions, respectively, in the image plane z_B.

The principal electrical and magnetic deflection planes are obtained in the form of an intersect of the optical axis with the tangent at $|w_U(z)|$ or $|w_I(z)|$ in the image plane z_B:

$$z_{AHU} = z_B - \frac{|w_{UB}|}{|w_U|'_B} = z_B - \frac{x_{UB}^2 + y_{UB}^2}{x_{UB}x'_{UB} + y_{UB}y'_{UB}} \tag{87a}$$

$$z_{AHI} = z_B - \frac{|w_{IB}|}{|w_I|'_B} = z_B - \frac{x_{IB}^2 + y_{IB}^2}{x_{IB}x'_{IB} + y_{IB}y'_{IB}} \tag{87b}$$

The intersection of the tangent with the optical axis is only secured by applying the tangent to the modular trajectories $|w_U|$ or $|w_I|$. For in the presence of a magnetic round lens field, or if the axial dipole functions P_1 or Q_1 are complex, the tangent at w_U or w_I will generally approach the optical axis as a skew ray. One may compute not only intersects (87a) and (87b), but also the resulting intersects for the projections in the x- and y-sections which are considered separately:

$$z_{AHUX} = z_B - x_{UB}/x'_{UB} \qquad z_{AHUY} = z_B - y_{UB}/y'_{UB} \tag{88a}$$

$$z_{AHIX} = z_B - x_{IB}/x'_{IB} \qquad z_{AHIY} = z_B - y_{IB}/y'_{IB} \tag{88b}$$

A comparison of the two values in line with Eq. (88a) or (88b) with the values obtained from Eq. (87a) or (87b) provides us with a measure for the ray skewness of the deflection trajectories in the image plane. We could, for example, also compare $\arg(w_{UB})$ with $\arg(w'_{UB})$, or $\arg(w_{IB})$ with $\arg(w'_{IB})$ as a measure of the ray skewness.

In e-beam lithography, the deflected beam must also be incident on the resist surface as perpendicular as possible. It is therefore usual (Goto & Soma, 1977) to define a vertical landing aberration as $K_V = w'_{DB}/w_{DB}$. One may calculate the two landing aberrations for electrical and magnetic deflections separately:

$$K_{VU} = w'_{UB}/w_{UB} \quad \text{and} \quad K_{VI} = w'_{IB}/w_{IB} \tag{89}$$

Kruit and Dubbeldam (1987) and Muray and Richardson (1987) use a VAIL for vertical landing of the PEs and magnetic collimation of the SEs simultaneously. The VAIL (Sturans & Pfeiffer, 1983), which represents a clever superposition of the magnetic lens and deflection field, has the maximum of $B(z)$ in the target plane and has already been successfully used for a number of years in e-beam writing. Vertical landing is also important for the vidicon, where an electron beam is decelerated to a few electronvolt scans and neutralizes the photoconductive layer.

3.3 Aberrations

The papers of Soma (1977), Chu and Munro (1982a), and Plies (1982a, 1982b) deal with the calculation of the aberrations of an electron optical compound system from any combination or superposition of magnetic and electrostatic lenses and deflection fields. They differ mainly in their approach and completeness when considering the third-order geometrical aberrations, the first-order chromatic aberrations, the alignment aberrations and dynamic correctors.

A finite object size is taken into account by Soma (1977) for the image aberrations, but only the total deflection in the image plane is considered, i.e. it is not separated into its electrostatic and magnetic deflection components. Soma only takes into account rotationally invariant deflection fields, but also explicitly considers dynamic fields for correcting the focus, two-fold astigmatism and distortion during beam deflection. However, Li (1983) states that unfortunately the image aberration expressions given by Soma (1977) are not correct in the presence of an electrostatic lens field since it appears that only the Wronski determinant of the purely magnetic lens field was used in the perturbation theory. Chu and Munro (1982a) have treated all cases of e-beam lithography currently relevant in practice: a Gaussian round beam with dual-channel deflection, a shaped beam with magnetic deflection and a shaped beam with electrostatic deflection.

The general case with a finite object size and magnetic and electrostatic deflection has been treated by the author (Plies, 1982a, 1982b). Dynamic focusing, dynamic stigmators with quadrupole and hexapole fields and a dynamic distortion correction are also taken into account there. However, the case of the general, dual-channel deflection system is not explicitly included. Formulae for the alignment aberrations of second rank are also given.

By variation of parameters, the first-order deflection trajectory was determined in Section 3.2. The same method can be used to transform trajectory equation (70a) with the sum $\Sigma \neq 0$ into an integral equation:

$$w = w_0 + \frac{w_\alpha}{\sqrt{\Phi_G}} \int_{z_G}^{z} \sqrt{\Phi}\,\Sigma\left(w, \overline{w}, w', \overline{w}', w'', \overline{w}'', z, U, \overline{U}, I, \overline{I}, \ldots\right)\overline{w}_\gamma \, dz$$
$$- \frac{w_\gamma}{\sqrt{\Phi_G}} \int_{z_G}^{z} \sqrt{\Phi}\,\Sigma\left(w, \overline{w}, w', \overline{w}', w'', \overline{w}'', z, U, \overline{U}, I, \overline{I}, \ldots\right)\overline{w}_\alpha \, dz \qquad (90)$$

where w_0 is the already known electron trajectory of the focusing and the first-order deflection. This non-linear integrodifferential equation of the second kind can be solved in the usual way by the method of successive approximation. The trajectory w_0 is substituted for w in Eq. (90) to obtain the first approximation. To calculate the third-order geometrical aberrations, the first-order chromatic aberrations and the alignment aberrations of second rank, no further iteration steps are required for electron optical systems with a straight axis. Therefore, the expression

$$\Delta w_B = w_B - w_{0B} = \frac{-w_{\gamma B}}{\sqrt{\Phi_G}} \int_{z_G}^{z_B} \sqrt{\Phi}\Sigma\left(w_0, \overline{w}_0, w_0', \overline{w}_0', w_0'', \overline{w}_0'', z, U, \overline{U}, I, \overline{I}, \ldots\right)\overline{w}_\alpha \mathrm{d}z \quad (91)$$

for the image aberration is obtained from Eq. (90) and is the difference between the trajectory obtained in this way and the trajectory w_0 in the Gaussian image plane $z = z_B$, taking $w_\alpha(z_B) = 0$ into account. Let us now return to the decomposition of Σ into non-linear perturbation terms as defined in Eq. (70c) and discuss the physical interpretation of the terms in the sum after making the substitution $w = w_0$.

$s_3 = s_3\ (w_0, \overline{w}_0, \ldots, z, \ldots)$ contains the axial functions Φ, B, P_1, Q_1, P_3, Q_3 and their derivatives and is a cubic in α, $\overline{\alpha}$, γ, $\overline{\gamma}$, U, $\overline{U}$, I, $\overline{I}$ and therefore determines the geometrical aberrations of third (Seidel) order. This extension of the usual Seidel order of aberrations with the inclusion of the beam deflection (U, I or beam deflection in the image plane) has already been justified by Glaser (1949, 1952) for electron optical systems with a straight axis. All further perturbation terms in Eq. (70c) cause chromatic aberrations (index c) or alignment aberrations (index J), or describe the contribution of dynamic correctors (index K). To classify them, we will use the concept of the rank of an aberration also used by Rose (1987). The rank of an aberration is defined as the sum of the (extended) order of the aberration and all the powers of other perturbations on which the aberration depends. Other perturbations may be, for example, the energy spread of the electrons, characterized by $e\Delta\Phi_c$ (semi half-width value). The perturbation term s_{2c} is linear in w, w', w'', U and linear in $\Delta\Phi_c$ and causes the chromatic aberrations of first order and second rank. The expression for s_{2c} is obtained by substituting $\Phi(z) + \Delta\Phi_c$ for the axial acceleration potential $\Phi(z)$ in the potential expansion.

The axial functions Φ, B, P_1, Q_1, P_3, and Q_3 describe an ideally aligned electron optical system comprising lenses and deflection elements with two planes of symmetry. Construction errors, misalignments, and external, stray fields will destroy the assumed symmetries. This can be taken into account by the perturbation potentials φ_Δ and ψ_Δ which possess multipole components $\varphi_{\nu\Delta}$ and $\psi_{\nu\Delta}$ of any order. At this juncture, we will not discuss the explicit relationship between the perturbation potentials and, say, the geometrical deviations. For further details of this topic, reference is made to publications such as those of Sturrock (1951), Glaser (1952), Glaser and Schiske (1953) and Hahn (1959a). Using the perturbation potentials φ_Δ and ψ_Δ, their multipole components $\varphi_{\nu\Delta}$ and $\psi_{\nu\Delta}$ and also the associated axial multipole functions $\Phi_{\nu\Delta}$ and $\Psi_{\nu\Delta}$, not only disturbances but also static alignment errors, stigmators, fine-focusing coils and distortion correctors can be taken into account. This is particularly useful for alignment aberration compensation. If an experiment has revealed a specific alignment aberration (e.g. elliptical distortion), then its aberration integral can be used to determine which element should be used to correct it (in this case quadrupole, $\Phi_{2\Delta}$ or $\Psi_{2\Delta}$) and to determine the best location for the element.

The perturbation term s_{3K} takes the contributions from the dynamic correctors into account and exhibits homogeneity of the third degree in w, $\overline{w}$, w', U, $\overline{U}$, I, $\overline{I}$. In the case of dynamic correctors, a distinction is made between the following:

- dynamic focusing (round lens fields quadratic in U, $\overline{U}$, I, $\overline{I}$, for correcting the deflection defocus, i.e. field curvature of deflection);
- dynamic two-fold stigmators (quadrupole fields quadratic in U, I, or $\overline{U}$, $\overline{I}$, for correcting two-fold deflection astigmatism);
- dynamic three-fold stigmators (hexapole fields linear in $\overline{U}$, $\overline{I}$, for correcting three-fold deflection astigmatism);
- dynamic distortion correctors (dipole fields, cubic in U, $\overline{U}$, I, $\overline{I}$, for correcting deflection distortion, i.e. for correcting the deflection nonlinearity).

In the case of e-beam testing and electron optical inspection, the beam deflections are usually not large enough to require dynamic correction for deflection aberrations. This only applies to e-beam writing, so the discussion of dynamic correctors will not be taken any further.

Geometrical Third-Order Aberrations

Let us now return to third-order geometrical aberrations which were mentioned briefly before. If we turn our attention to the s_3 component in the sum Σ of all perturbation terms given in Eq. (91) and make the substitution for $w_0, \overline{w}_0, w_0', \ldots$ described by Eq. (86) and, in the integral, collect together terms of the sum with the same dependence on the trajectory parameters $\alpha, \overline{\alpha}, \gamma, \overline{\gamma}, U, \overline{U}$, and $I, \overline{I}$, then Eq. (91) can be decomposed into individual aberration integrals. As can be seen from Table 1, there are 56 aberration constants (aberration integrals) which are, in general, complex. Six of these constants are related to pure image aberrations, 26 to pure deflection aberrations and 24 to mixed image deflection aberrations. Of the 56 aberration constants in Table 1, 40 describe rotationally invariant aberrations, i.e. aberrations which are invariant to combined azimuth rotation of the deflection signals U, I and the coordinate system.

To determine the aberration constants, 56 different, complicated integrals must be calculated using numerical analysis. The best way of doing this is to reduce the large numbers of aberration integrals to a small number of general types, the original integrals being obtained by changing the arguments. This method has been used by Munro and Wittels (1977), Plies and Typke (1978), Rose (1978), Chu and Munro (1982a), and Plies (1982b) and has many advantages when subprograms are used to perform the numerical calculations.

It should be noted that a distinction must be made between objective and projector modes, as is done with the cardinal elements, when aberration integrals are calculated. This means that in the asymptotic or projector mode the integral must be calculated over the whole field region, even if there is a real image in this region. Hawkes (1989) gives the structure of the asymptotic aberration coefficients of combined focusing and deflection systems and uses polynomials in the reciprocal magnification.

On the basis of the eikonal method, the following relations are found between the aberration constants (Plies, 1982b):

Spherical aberration:

$$K_{\alpha\alpha\overline{\alpha}} = \overline{K}_{\alpha\alpha\overline{\alpha}} \tag{92a}$$

Field curvature:

$$\left.\begin{array}{lll} K_{\alpha\gamma\overline{\gamma}} = \overline{K}_{\alpha\gamma\overline{\gamma}} & K_{\alpha\gamma\overline{U}} = \overline{K}_{U\alpha\overline{\gamma}} & K_{\alpha\gamma\overline{I}} = \overline{K}_{I\alpha\overline{\gamma}} \\ K_{U\alpha\overline{U}} = \overline{K}_{U\alpha\overline{U}} & K_{I\alpha\overline{I}} = \overline{K}_{I\alpha\overline{I}} & K_{U\alpha\overline{I}} = \overline{K}_{I\alpha\overline{U}} \end{array}\right\} \tag{92b}$$

Coma:

$$K_{\alpha\gamma\overline{\alpha}} = 2\overline{K}_{\alpha\alpha\overline{\gamma}} \quad K_{U\alpha\overline{\alpha}} = 2\overline{K}_{\alpha\alpha\overline{U}} \quad K_{I\alpha\overline{\alpha}} = 2\overline{K}_{\alpha\alpha\overline{I}} \tag{92c}$$

Table 1 Geometrical third-order aberrations

	Type			Name		Origin
	Pure image aberrations	Mixed image–deflection aberrations	Pure deflection aberrations			
$\frac{\Delta w_{B}^{(3)}}{-w\gamma_{B}} =$	$K_{aa\bar{a}}a^2\bar{a}$			Spherical aberration	Rotationally invariant aberrations	Round lenses and/or dipole components of deflectors
	$+K_{a\gamma\bar{a}}a\bar{a}\gamma$		$+K_{Ua\bar{a}}a\bar{a}U + K_{Ia\bar{a}}a\bar{a}I$	Coma length		
	$+K_{aa\bar{\gamma}}a^2\bar{\gamma}$		$+K_{aa\overline{U}}a^2\overline{U} + K_{aa\bar{I}}a^2\bar{I}$	Coma radius		
	$+K_{a\gamma\bar{\gamma}}a\gamma\bar{\gamma}$	$+K_{a\gamma\overline{U}}a\gamma\overline{U} + K_{a\gamma\bar{I}}a\gamma\bar{I} + K_{Ua\bar{\gamma}}a\bar{\gamma}U + K_{Ia\bar{\gamma}}a\bar{\gamma}I$	$+K_{Ua\overline{U}}aU\overline{U} + K_{Ua\bar{I}}aU\bar{I} + K_{Ia\overline{U}}a\overline{U}I + K_{Ia\bar{I}}aI\bar{I}$	Field curvature		
	$+K_{\gamma\gamma\bar{a}}\bar{a}\gamma^2$	$+K_{U\gamma\bar{a}}\bar{a}\gamma U + K_{I\gamma\bar{a}}\bar{a}\gamma I$	$+K_{UU\bar{a}}\bar{a}U^2 + K_{UI\bar{a}}\bar{a}UI + K_{II\bar{a}}\bar{a}I^2$	Two-fold astigmatism		
	$+K_{\gamma\gamma\bar{\gamma}}\gamma^2\bar{\gamma}$	$+K_{\gamma\gamma\overline{U}}\gamma^2\overline{U} + K_{\gamma\gamma\bar{I}}\gamma^2\bar{I} + K_{U\gamma\bar{\gamma}}\gamma\bar{\gamma}U + K_{I\gamma\bar{\gamma}}\gamma\bar{\gamma}I$ $+K_{U\gamma\overline{U}}\gamma U\overline{U} + K_{U\gamma\bar{I}}\gamma U\bar{I} + K_{I\gamma\overline{U}}\gamma\overline{U}I + K_{I\gamma\bar{I}}\gamma I\bar{I}$ $+K_{UU\bar{\gamma}}\bar{\gamma}U^2 + K_{UI\bar{\gamma}}\bar{\gamma}UI + K_{II\bar{\gamma}}\bar{\gamma}I^2$	$+K_{UU\overline{U}}U^2\overline{U} + K_{UU\bar{I}}U^2\bar{I} + K_{UI\overline{U}}U\overline{U}I$ $+K_{UI\bar{I}}UI\bar{I} + K_{II\overline{U}}\overline{U}I^2 + K_{II\bar{I}}I^2\bar{I}$	Distortion		
		$+K_{\overline{a\gamma}\overline{U}}\overline{a\gamma}\overline{U} + K_{\overline{a\gamma}\bar{I}}\overline{a\gamma}\bar{I}$	$+K_{\overline{U}\bar{a}\overline{U}}\bar{a}\overline{U}^2 + K_{\overline{U}\bar{a}\bar{I}}\bar{a}\overline{UI} + K_{\bar{I}\bar{a}\bar{I}}\bar{a}\bar{I}^2$	Two-fold astigmatism	Hexapole aberrations	Hexapole components of deflectors
			$+K_{\overline{aa}\overline{U}}\bar{a}^2\overline{U} + K_{\overline{aa}\bar{I}}\bar{a}^2\bar{I}$	Three-fold astigmatism		
		$+K_{\overline{\gamma\gamma}\overline{U}}\bar{\gamma}^2\overline{U} + K_{\overline{\gamma\gamma}\bar{I}}\bar{\gamma}^2\bar{I}$ $+K_{\overline{U\gamma U}}\bar{\gamma}\overline{U}^2 + K_{\overline{U\gamma I}}\bar{\gamma}\overline{UI} + K_{\overline{I\gamma I}}\bar{\gamma}\bar{I}^2$	$+K_{\overline{UUU}}\overline{U}^3 + K_{\overline{III}}\bar{I}^3$ $+K_{\overline{UUI}}\overline{U}^2\bar{I} + K_{\overline{IIU}}\overline{U}\bar{I}^2$	Distortion		

By this means, the number of rotation-invariant third-order image aberration constants is reduced from 40 complexes = 80 real constants to $64 = 4^3$ real constants. Eqs. (92a)–(92c), therefore, represent an excellent test, first for computation errors in the trajectory method, secondly for programming errors and thirdly for numerical accuracy in the image aberration calculation. That there are too few sample points in the axial fields, and thus too few for calculating the trajectories and image aberrations, becomes immediately apparent from minor or major departures from relations (92a)–(92c).

The image aberration constants from Table 1 are also designated as image aberration constants with respect to the object plane because $\Delta w_B^{(3)}$ was divided by the complex lateral magnification $w_{\gamma B}$ (with a minus sign due to image inversion) and because the object side trajectory parameters α and γ were used. One may also calculate the image aberration constants with respect to the image plane, if required. In this case, the expression for image aberration contains the following trajectory parameters: image-side aperture α_B, lateral dimension of the image γ_B, electrical deflection in the image plane, W_e, and magnetic deflection in the image plane, W_m. These image side trajectory parameters are related to the previously used trajectory parameters in the following way:

$$\alpha_B = \alpha w'_{\alpha B} \qquad \gamma_B = \gamma w_{\gamma B} \qquad W_e = U w_{UB} \qquad W_m = I w_{IB} \tag{93}$$

Eq. (93) shows once again that assuming U and I are small for the perturbation calculation is equivalent to the statement that the deflections W_e and W_m are small (small compared with the length of the deflection field).

It should be noted that the terms used in Table 1 are by no means generally accepted. Munro (1974) and Chu and Munro (1982a) called the aberration referred to here as hexapole aberrations four-fold aberrations because they are the result of the four-fold symmetry (rotation through $90° = 360°/4$) of the deflection fields which can be seen from Eq. (68a).

Great differences occur under the headings of single parts of four-fold aberrations or hexapole aberrations. The four-fold coma radius of Munro (1974) or the four-fold coma of Chu and Munro (1982a) are referred to in Table 1 as three-fold astigmatism, the epithet "three-fold" designating the aperture cardinality (a 360° excursion round the complex aperture $\alpha = |\alpha| e^{i\vartheta}$ generates a three-fold aberration figure in a plane $z = z_B \pm \Delta z$). The designation "coma" is somewhat unfortunate because the aberration does not produce the characteristic coma shape (Born, 1933; Scherzer, 1937; Glaser, 1952) and it can be corrected using a dynamic three-fold stigmator (dynamic hexapole field). Glaser (1952) uses the term "deflection coma". In terms of our terminology, it is a combination of the deflection coma ($K_{U\alpha\bar{\alpha}}$, $K_{I\alpha\bar{\alpha}}$, $K_{\alpha\alpha\overline{U}}$, $K_{\alpha\alpha\overline{I}}$) and the three-fold deflection astigmatism ($K_{\overline{\alpha\alpha}\overline{U}}$, $K_{\overline{\alpha\alpha}\overline{I}}$), which is why Glaser's coma circle is not a circle but an ellipse. The terms used here, by Plies (1982b), and by Plies and Elstner (1989a), would seem to be adequate when judged from the point of view of the cause, the form and the effect of each term in the sum, and can trace their origins back to the lectures and suggestions of O. Scherzer and H. Rose. Tables with aberration figures which can be very useful for diagnosing aberrations have been published by Wendt (1939, 1954), Plies (1973), and Bernhard (1980), for example.

Chromatic and Alignment Aberrations of Second Rank

To calculate chromatic aberrations of second rank (first order), the procedure is formally exactly the same as the calculation of third-order geometrical aberrations in the last section and only takes into account the perturbation term s_{2c} which has been mentioned previously instead of the trajectory perturbation s_3. The result is an expression for the chromatic aberration referred to the object plane:

$$\frac{\Delta w_{\mathrm{Bc}}^{(2)}}{-w_{\gamma\mathrm{B}}} = (K_{c\alpha}\alpha + K_{c\gamma}\gamma + K_{cU}U + K_{cI}I)\frac{\Delta\Phi_{\mathrm{c}}}{\Phi_{\mathrm{G}}} \tag{94}$$

The four aberration constants may be reduced to a general integral which was specified in Plies (1982b). $K_{c\alpha}$ describes the axial chromatic aberration, which is the most important aberration in low-voltage electron optics, and must always be real. $K_{c\gamma}$ describes the off-axial (or transverse) chromatic aberration, and K_{cU} and K_{cI} the chromatic aberrations of the electrical and magnetic deflections, respectively.

By considering the special case where deflection takes place after focusing without any overlap of the deflection and focusing fields, the pure chromatic deflection aberration with respect to the image plane can be reduced to the simple form below:

$$\Delta w_{\mathrm{Bc}}^{(2)}|_{\alpha=\gamma=0} = -\left(W_{\mathrm{e}} + \frac{W_{\mathrm{m}}}{2}\right)\frac{\Delta\Phi_{\mathrm{c}}}{\Phi_{\mathrm{B}}} \tag{95}$$

In this case, therefore, the magnetic chromatic deflection aberration is half as great as the electric chromatic deflection aberration and both aberrations do not depend on the axial variation of the magnetic and electrical deflection field strengths. If the magnetic deflection is chosen to be twice as great as the electrical deflection and in the opposite direction, the chromatic deflection aberration vanishes in this special case, but the total deflection remains finite: $W = W_e + W_m = -W_e = W_m/2$.

In practical electron optical systems, aberrations can be caused by parasitic fields that arise due to misalignments. It is certainly possible that these aberrations will exceed the aberration of the wanted fields. These parasitic fields, as well as static fine focusing, alignment, stigmator and distortion correction fields, are included in the trajectory perturbations s_{1J}, s_{2J}, and s_{2Jc} and give the alignment aberrations of first and second rank in Table 2. The aberration integrals of these alignment aberrations were given by Plies (1982b).

The coefficients of the alignment aberrations in Table 2 have, in addition to the indexing of the aberration coefficients of the aligned systems, one or two Δ as an index. This expresses the linear or the quadratic dependence of the aberration integrals on the perturbation potentials. The coefficient of defocussing $K_{\Delta\alpha}$ must always be real while the other coefficients are, in general, complex.

Electron Probe Size

Care must be taken when calculating the lateral dimension of the electron probe from all the individual aberrations. For scanning systems with a Gaussian round beam, it is common practice to consider the square root of the sum of the squares of the first-order probe diameter ($= 2\gamma_B$) of the diameter of the Airy disc ($= 1.22\lambda_B/\alpha_B$, where λ_B is the wavelength at the image side) and of all the different aberrations for a given aperture α_B, energy spread $e\Delta\Phi_c$ and deflection positions. (Aberrations depending on γ_B have to be considered only for scanning systems with shaped-beam and fixed-beam imaging systems.) This method would be correct if all contributions—particularly the contributions from individual aberrations—had a Gaussian distribution possessing rotational symmetry. For the undeflected electron probe ($W_e = W_m = 0$), rotational symmetry at least is present so that the quadratic superposition of the diameter of first order, diffraction, spherical aberration and axial chromatic aberration provides an acceptable worst case estimate. It is, however, better to use the wave optical aberration function, as defined in Born and Wolf (1985). Shao and Crewe (1988) have, in this way, obtained a rather complicated expression for the optimal probe diameter when the spherical and axial chromatic aberration both make

Table 2 Alignment and chromatic aberrations of first and second rank

Aberration expressions		Name	Order	Rank	Type
$\frac{\Delta w_{BJ}^{(1)}}{-w_{\gamma B}} = K_{\Delta}$		Beam displacement	0	1	Alignment aberrations
$\frac{\Delta w_{BJ}^{(2)}}{-w_{\gamma B}} = K_{\Delta\Delta}$			0	2	
$+ K_{\Delta a} a$		Defocus	1	2	
$+ K_{\Delta \bar{a}} \bar{a}$		Two-fold astigmatism	1	2	
$+ K_{\Delta \gamma} \gamma$	$+ K_{\Delta U} U + K_{\Delta I} I$	Distortion	1	2	
$+ K_{\Delta \bar{\gamma}} \bar{\gamma}$	$+ K_{\Delta \overline{U}} \overline{U} + K_{\Delta \bar{I}} \bar{I}$				
$+ K_{\Delta C} \Delta\Phi_C / \Phi_G$		Dispersion	0	2	Alignment aberrations / Chromatic aberrations
$\frac{\Delta w_{BC}^{(2)}}{-w_{\gamma B}} = K_{Ca} a \Delta\Phi_C / \Phi_G$		Axial chromatic aberration	1	2	Chromatic aberrations
$+ K_{C\gamma} \gamma \Delta\Phi_C / \Phi_G$		Off-axial chromatic aberration	1	2	
	$+(K_{CU} U + K_{CI} I) \Delta\Phi_C / \Phi_G$	Chromatic deflection aberration	1	2	

From Plies (1982b).

a substantial contribution to the probe diameter, although the first-order probe diameter is negligible. If the acceleration voltages are very low, the contribution of spherical aberration can also be neglected and a relatively simple formula,

$$d_{\mathrm{B}} = \sqrt{2\lambda_{\mathrm{B}} K_{c\alpha,\mathrm{B}} \Delta\Phi_c / \Phi_{\mathrm{B}}} = 8.8 \text{ nm} \sqrt{\frac{(K_{c\alpha,\mathrm{B}}/\mathrm{mm})(\Delta\Phi_c/\mathrm{V})}{(\Phi_{\mathrm{B}}/\mathrm{kV})^{3/2}}} \tag{96a}$$

is derived for the minimum obtainable probe diameter d_{B}. The optimal aperture that should be chosen is

$$\alpha_{\mathrm{B}} = \sqrt{\frac{3}{4} \frac{\lambda_{\mathrm{B}} \Phi_{\mathrm{B}}}{K_{c\alpha,\mathrm{B}} \Delta\Phi_c}} = 5.4 \text{ mrad} \sqrt{\frac{\sqrt{\Phi_{\mathrm{B}}/\mathrm{kV}}}{(K_{c\alpha,\mathrm{B}}/\mathrm{mm})(\Delta\Phi_c/\mathrm{V})}} \tag{96b}$$

To make these formulae clearer, we will make the constant for the axial chromatic aberration referred to the image plane $K_{c\alpha,\mathrm{B}} = 4$ mm, the acceleration potential in the image plane $\Phi_{\mathrm{B}} = 0.5$ kV and half the energy spread of the electrons $\Delta\Phi_{\mathrm{C}} = 0.25$ V. Then Eq. (96a) gives a value of 15 nm for d_{B} and Eq. (96b) gives a value of 4.5 mrad for the optimal aperture in the image plane, α_{B}. We will see that values of that kind can be reached by compound magnetic electrostatic retarding lenses without the object under examination having to lie in the lens field.

Both the axial aberrations, the spherical aberration and the axial chromatic aberration, are, as Scherzer's theorem shows (Scherzer, 1936), unavoidable in the case of static, rotationally symmetrical lenses without space charge. Tables and/or curves of the aberration constants for purely magnetic lenses will be found in Dugas et al. (1961), El-Kareh and El-Kareh (1970b), and Hawkes (1982), for example, and for the purely electrostatic lens in El-Kareh and El-Kareh (1970b), Harting and Read (1976), and Szilagyi (1988), for example. For spatially superposed electrostatic/magnetic lenses there is no comparable corpus of data. The formula below for converting the axial aberration constants from the object position z_{G} to the image position z_{B} is valid for any type of round lens:

$$K_{\alpha\alpha\bar{\alpha},\mathrm{B}} = \left(\frac{\Phi_{\mathrm{B}}}{\Phi_{\mathrm{G}}}\right)^{3/2} V^4 K_{\alpha\alpha\bar{\alpha},\mathrm{G}} \tag{97a}$$

$$K_{c\alpha,\mathrm{B}} = \left(\frac{\Phi_{\mathrm{B}}}{\Phi_{\mathrm{G}}}\right)^{3/2} V^2 K_{c\alpha,\mathrm{G}} \tag{97b}$$

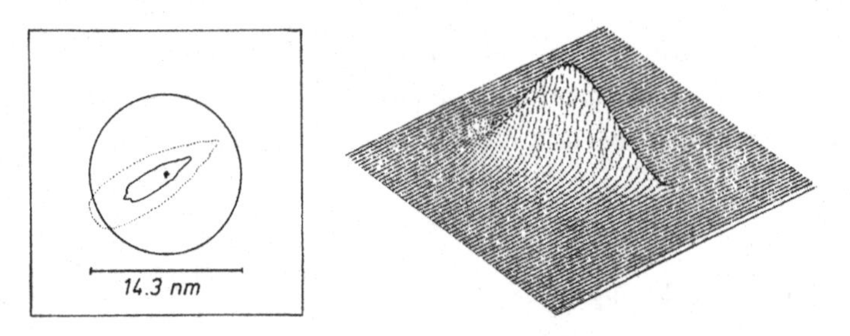

Figure 15 Intensity distribution for a deflected point spot, caused by coma, spherical aberration and chromatic aberrations; wave optical result. Broken line, 10% of maximum intensity; solid line, 50% of maximum intensity; the circle represents the spot size obtained from the square of the sum of aberrations. *From Kern (1979).*

These equations are obtained by using Helmholtz's theorem (82) and, apart from conversions, are an important aid for understanding low-voltage immersion lenses. The reader will also recall that for the immersion lens the asymptotic focal lengths (projective focal lengths) satisfy the equation $f_{BA}/f_{GA} = (\Phi_B/\Phi_G)^{1/2}$.

Because the overall effect of the aberrations of a deflected beam is extremely complicated, Chu and Munro (1982a) have successfully plotted the shape of the aberrated spot in the form of a "spot diagram"; this is generated from the set of computed aberration constants by taking a large number of rays which are uniformly distributed in the aperture plane, and plotting their aberrated positions in the image plane. In a similar way, Kern (1979) has calculated the intensity distribution for a defocussed deflected point spot, caused by astigmatism, coma, spherical aberration and chromatic aberrations. He assumed the intensity was proportional to the density of intersects of trajectories with the image plane and calculated these intersects by evaluating the expression for aberration using a large number of trajectories, representative of the beam. Kern (1979), however, has also performed a rigorous wave optical calculation to obtain the intensity distribution of a deflected point spot whilst taking coma, spherical aberration and chromatic aberration into account; the result is seen in Fig. 15. To do this, the intensity distribution was calculated for various energies (from the energy spread of the electrons) and then integrated over the beam energy, a Gaussian energy distribution being used for the weighting function.

Fig. 16A illustrates the principle of a compound spectrometer objective lens for e-beam testing (Plies and Schweizer, 1987; Plies and Elstner, 1989b). Fig. 16B shows the light optical counterpart of the spectrometer

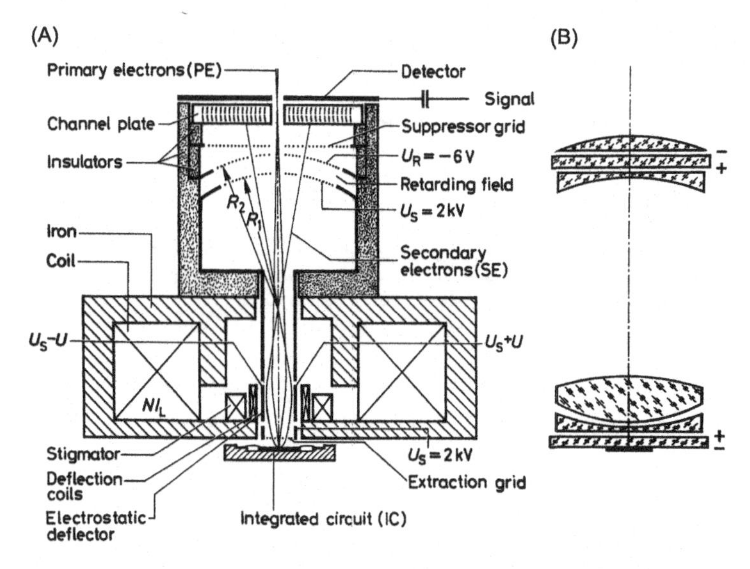

Figure 16 (A) Schematic diagram of a spectrometer objective lens with PE and SE rays, whose scales are radially enlarged by a factor of 10 and 3, respectively. $U_S = 2$ kV = extraction voltage; $U_R = -6$ V = retarding voltage; U = deflection voltage; $NI_L = 458$ At = lens excitation for a final PE energy of 700 eV at the IC; grid radii $R_1 = 30$ mm and $R_2 = 34$ mm. (B) Light optical counterpart of the spectrometer objective lens relative to the PE focusing. The thick convex lens corresponds to the magnetic objective lens. All other lenses and parallel plates correspond to the electrostatic fields. Each electrostatic lens is formed by the axial grid mesh and the existing acceleration or retarding fields on both sides of the mesh. *From Plies and Schweizer (1987).*

objective relative to the PE focusing. At this point, we are only concerned with PE probe forming. Fig. 17A shows the axial functions of the round lens and deflection fields. From this, in essence using the method described previously, the fundamental trajectories of the focusing and the deflection shown in Fig. 17B were calculated, as were the aberration coefficients with respect to the image plane, of which only the four most important (spherical aberration, axial chromatic aberration, electrical and magnetic deflection coma) are given as numerical values. The coma coefficients $K_{E\alpha\overline{\alpha}}$ and $K_{M\alpha\overline{\alpha}}$ have the indices E and M which indicate the linear dependence of the aberrations on the electrical deflection $W_e = Uw_{UB}$ or the magnetic deflection $W_m = Iw_{IB}$ in the image plane. The probe diameters, which are also given in Fig. 17B, are calculated using the root of the sum of the squares of the individual aberrations and the first-order diameter ($2\gamma_B$).

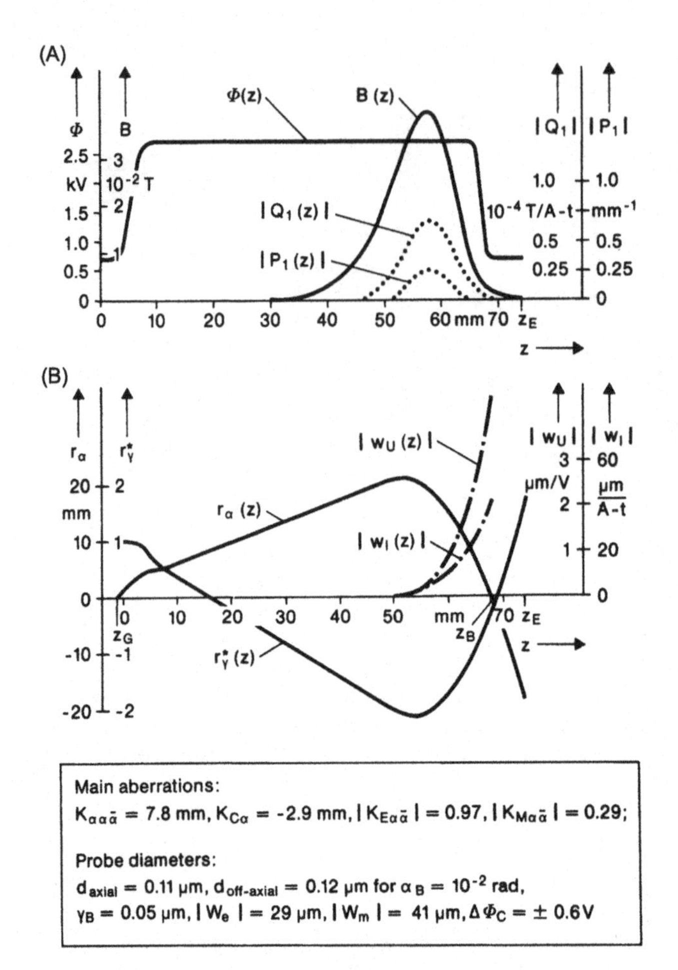

Main aberrations:
$K_{\alpha\alpha\bar{\alpha}}$ = 7.8 mm, $K_{C\alpha}$ = -2.9 mm, $|K_{E\alpha\bar{\alpha}}|$ = 0.97, $|K_{M\alpha\bar{\alpha}}|$ = 0.29;

Probe diameters:
d_{axial} = 0.11 μm, $d_{off\text{-}axial}$ = 0.12 μm for $\alpha_B = 10^{-2}$ rad,
γ_B = 0.05 μm, $|W_e|$ = 29 μm, $|W_m|$ = 41 μm, $\Delta\Phi_C = \pm$ 0.6V

Figure 17 (A) Axial potential and field distributions $\Phi(z)$, $B(z)$, $P_1(z)$, and $Q_1(z)$ with (B) electron optical performance data and the fundamental trajectories $r_\alpha(z)$, $r^*_\gamma(z)$, $w_U(z)$, and $w_I(z)$ of the spectrometer objective lens of Fig. 16. The aberration constants are referred to the image plane z_B and the lateral magnification is $V = -0.34$. *From Plies and Elstner (1989b).*

3.4 Ray Tracing

For the perturbation calculations in Sections 3.2 and 3.3, it was assumed that both the distance of the trajectory $w(z)$ from the axis and the trajectory gradient $w'(z)$ were small. The second condition $|w'| \ll 1$ is not met by electron guns because emission angles of up to 90° can occur. This is why ray tracing is usually preferred for electron optical calculations on guns, i.e. the numerical integration of the equation of motion (53), or an exact trajectory equation, e.g. Eq. (65) or (67). This is, however, not obligatory if it is only the condition $|w'| \ll 1$ that is not satisfied. This is because Hahn (1958a) brought the general differential equation for the trajectory curve in an electrostatic field into a form which could be solved approximately without any restrictions on the slope of the trajectory curves having to be met. The procedure was originated by Recknagel (1941) and is based on transferring the role played by the small quantity, which in the case of conventional lenses is usually the trajectory inclination, to the exit velocity. Hahn (1958a) was then able to state a fundamental trajectory system, cardinal elements and aberration coefficients of the third order and to evaluate them for special electrostatic immersion objectives (Hahn, 1958b, 1959b).

The trend in electron optical calculations is moving away from perturbation methods to numerical integration of the equation of motion or an exact trajectory equation. For certain applications (e.g. detectors) visual assessment of the plotted trajectories is all that is required, in other cases, current density profiles can be calculated from the points where the trajectories pass through the image plane. To a greater extent, even aberration constants are being calculated using the points at which the trajectory passes through the image plane, even when aberration integrals for numerical evaluation are available from the perturbation method. This is only possible because modern computers are powerful enough to calculate the spatial field distribution with sufficient accuracy, because the methods of numerical trajectory integration are sufficiently accurate and stable, and because the methods for determining the aberration constants from a number of intersection points have been improved (Kasper, 1985). A further reason is that, in some cases, limiting the aberrations to the third order gives inaccurate results. However, taking perturbation theory to fifth-order aberrations leads to an impenetrable thicket of aberration integrals (including double integrals and products of integrals) which are unsuitable for determining current density profiles in electron probes. As numerical ray tracing gives little physical insight into ways of improving imaging characteristics, third-order aberration integrals, which often suggest such rules, will continue to play an important role even if they have to share the stage with numerical ray tracing.

Some General Remarks on Initial Value Problems in Ordinary Differential Equations

Although the equation of motion and all the different forms of the differential trajectory equation are of the second order, they, and the initial conditions, can nonetheless be reduced to a system of ordinary first-order differential equations by means of a suitable substitution. This system can then be solved using the standard methods of numerical analysis. These numerical methods can be classified as follows:

- One-step methods (methods of Euler–Cauchy, Heun, Runge–Kutta, and Runge–Kutta–Fehlberg);
- multistep methods (e.g. methods of Adams–Bashforth and Adams–Moulton);
- extrapolation algorithm (method of Bulirsch–Stoer–Gragg).

To calculate the next approximate value, one-step methods use only one previous value. Multistep methods use at least two previous values to calculate the next value. The extrapolation algorithm is a method analogous to Romberg's quadrature method (Romberg, 1955) which is used to solve initial value problems associated with ordinary differential equations. The predictor–corrector methods are a special class of one-step and multistep methods. These methods first of all find an approximation using a one-step or multistep method. This value or the algorithm used to obtain it is referred to as the predictor. The predictor is then improved by using a corrector. The methods of Heun and Adams–Moulton are predictor–corrector methods.

For more information on the methods listed above by way of classification, the reader is referred to the literature, e.g. the textbooks by Collatz (1960), Berezin and Zhidkov (1965), Zurmühl (1965), Jordan-Engeln and Reutter (1972), Ralston (1972), Hamming (1973), and Stoer and Bulirsch (1990). The literature referred to above also deals with questions of local and global errors, discretization and truncation errors, and the consistency, convergence and stability of the methods. Information on step width control and starting algorithms (for multistep methods) will also be found there. Jordan-Engeln and Reutter (1972), Enright and Hull (1976), and Stoer and Bulirsch (1990) also deal with criteria for selecting a particular method and give comparisons of results. A short résumé will now be given.

If computation time and accuracy are considered, the classical Runge–Kutta method (explicit fourth-order method) should certainly be preferred to the methods of Euler–Cauchy and Heun. It is a good idea to use the method of Adams–Bashforth as a predictor for the Adams–Moulton method. When the three methods—Runge–Kutta, Adams–Moulton, and Bulirsch–Stoer–Gragg—are compared, there is no one method that is clearly better than the others. The comparison depends on a multitude of factors, including the differential equation itself. If the right-hand side of the differential equation can be calculated reasonably efficiently, the extrapolation algorithm of Bulirsch–Stoer–Gragg is the best method. This applies to electron optics if the calculation of the required potential and field values is relatively efficient, e.g. relatively simple analytic formula or a moderate number of equivalent charges when the charge simulation method is used. If, on the other hand, the calculation of the right-hand side (field), which is used repeatedly, requires excessive effort, the predictor–corrector method of Adams–Moulton should be used. The Runge–Kutta methods should only be preferred when computation effort is reasonable and the accuracy requirements are none too stringent ($\approx 10^{-3}$). As far as boundary value problems are concerned, the reader should consult the textbooks that have already been mentioned as well as the review article by Walsh (1977).

Numerical Methods of Electron Ray Tracing

Hawkes (1980) provides a list of typical electron optical applications where Runge–Kutta routines and predictor–corrector methods can be used successfully (see this reference for more information). Only three further selected applications as examples illustrating the use of the standard algorithms of numerical analysis will be mentioned; we will then turn to special numerical procedures which have been proposed for use in electron optics. Van Hoof (1980) has successfully used the method of rational extrapolation described by Bulirsch and Stoer (1966) to trace electron rays far from the paraxial region of electrostatic systems with rotational symmetry. He used time as the independent variable and employed the CDM for field and potential calculations. Schönecker et al. (1986) obtained the SE trajectories in the microfield of three interconnection lines shown in Fig. 11B by solving the equation of motion numerically using the Runge–Kutta method. Plies and Schweizer (1987) have used a modified high-order predictor–corrector algorithm, which was derived from Zurmühl (1965), to obtain a numerical solution for the equation of motion of the secondary electrons in the compound spectrometer shown in Fig. 16. Fig. 18 shows the plots of the SE trajectories. The trajectories are projected onto the x–y plane in Fig. 18A, and Fig. 18B shows the radial coordinate of the trajectories along the optical axis. Both plots, $y(x)$ and $r(z)$, were obtained from the time-dependent trajectory coordinates $x(t)$, $y(t)$, and $z(t)$. The three examples which have been selected show that various standard numerical methods can be successful and that selecting the time t as the trajectory parameter is, in general, not fraught with drawbacks as is often suggested. As far as the last example is concerned (Figs. 16 and 18), it should be noted that the numerical calculation is performed so that each mesh grid passed (treated as

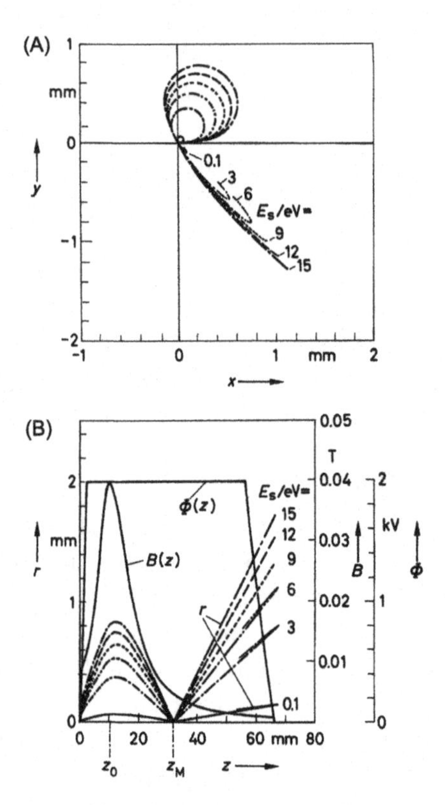

Figure 18 SE trajectories resulting from the combined influence of the electrostatic and magnetic fields inside the spectrometer objective. Final PE beam voltage of 700 V at the IC, lens excitation $NI_L = 458$ At. (A) Projection of the SE trajectories in the x–y plane perpendicular to the optical axis. (B) Radius r of the SE trajectories versus distance z along the optical axis. $\Phi(z) = \varphi(0, z)$ axial electrostatic potential with respect to the grounded device under test at $z = 0$; $B(z) = B_z(0, z)$, axial magnetic flux density; α_s, start aperture of the SEs with respect to the optical axis; E_s, start energy of the SEs. In these diagrams, $\alpha_s = 60°$ is fixed and E_s is varied. The SEs with $E_s = 0.1$, 3, and 6 eV are reflected for a retarding voltage of $U_R = -6$ V. The radii of the spherical grids of the spherical retarding field were chosen here as $R_1 = 30$ mm and $R_2 = 34$ mm (see also Fig. 16). *From Plies and Schweizer (1987).*

an electron-transmissive thin conductive foil; see kinks in $\Phi(z)$ in Fig. 18B) is struck out. To do this, a special shooting method was programmed and the accuracy of numerical ray tracing was checked using the conservation of energy and the conservation of the axial component of the canonical angular momentum (Eqs. (54) and (56)).

A still more accurate version of Hamming's modified predictor–corrector method (Hamming, 1959) has been developed by Kasper (1982,

1985). In this method, called "Hamming's modified predictor–corrector method in difference form" (HPCD), the accumulation of rounding errors is minimized by the use of a new numerically stable predictor formula, which has practically the same discretization error as Hamming's classical formula. Kasper calculates the first three sets of values (of the coordinates and velocities) by means of Gill's Runge–Kutta methods. Kasper claims that the HPCD is much faster than any Runge–Kutta routine and even superior to the extrapolation method described by Bulirsch and Stoer (1966), if, in the latter case, the field is obtained by interpolation in mesh grids. According to Kasper, the HPCD works well in this case, if Hermite interpolations are used. (See Kasper, 1982, 1985 for the HPCD formula system.)

Hahn (1985) has developed a special method for accurately calculating the trajectories in electron optical systems with a straight optical axis and deserves special attention because, from the point of view of electron optics, it is superior in many ways to general methods. Hahn starts with a second-order differential equation for the complex trajectory $w(z) = x(z) + iy(z)$ and introduces three new variables (one complex and two real) instead of w and z. By doing this, and by splitting up the expression for the field in a suitable way, he obtains two linear differential equations (one for the complex and one for the real variables) which are only weakly coupled via the higher-order field components and which can be solved using iteration. With the zeroth approximation, the trajectory lies in the optical axis, the first approximation being in the Gaussian region. With the second approximation, third-order aberrations arise but they are not the same as the image aberrations of third Seidelian order mentioned above. By repeating the iteration many times, quasi-accurate trajectories in the prescribed region are obtained. Specially developed integration formulae of the sixth order were used for this purpose. Apart from a straight axis, Hahn only assumes that the fields are constant with time and that space charge is absent. This means that complete calculations can be made on round lenses and deflection systems and even multipole optics and cylindrical lenses.

Numerical Solution of the Paraxial Ray Equation. In the well-known program package written by Munro (1973), a standard Runge–Kutta routine is used for solving the paraxial ray equation (73) in the case of purely magnetic and purely electrostatic lenses. The essential features of the paraxial ray equation (73) with $\Phi' = 0$ (purely magnetic case) and the paraxial ray equation (74) are their linearity and the absence of terms containing first derivatives of the trajectory coordinates r or R, respectively. For this type of ordinary differential equation, Numerov's method (Numerov, 1933) is usually recommended and used (e.g. see Moses, 1973; Kasper, 1982; Szilagyi, 1988; Plies & Elstner, 1989a). If we designate the step width by h, Numerov's difference equation for differential equation (74) is

$$R_{n+1} = \left\{\left(2 - \frac{5h^2}{6} f_n\right) R_n - \left(1 + \frac{h^2}{12} f_{n-1}\right) R_{n-1}\right\} \left(1 + \frac{h^2}{12} f_{n+1}\right)^{-1} \tag{98a}$$

with

$$f = \frac{3}{16} \frac{\Phi'^2}{\Phi^2} + \frac{e}{8m\Phi} B^2 \tag{98b}$$

This difference equation is very simple and numerically stable. Numerov's method is faster and more accurate than the standard Runge–Kutta method (the former has an intrinsic truncation error of the sixth order, while it is of the fifth order for the latter method). Runge–Kutta methods have, however, the advantage of being generally applicable. Since Numerov's method, as formulated by Eq. (98a), is a two-step method, a second sample point R_1 is required to continue with Eq. (98a), in addition to the well-known initial values R_0 and R'_0 to deal with the initial value problem. Plies and Elstner (1989a) used Heun's method for this initial step, whereas Kasper employed a more extensive start procedure whose error is, however, also only of the sixth order. After this start procedure has been performed, Eq. (98a) can be used successively to find a numerical solution of Eq. (74). This procedure has the disadvantage of using a constant step size h. However, the numerical integration can be stopped when necessary, h changed abruptly and Kasper's highly accurate start procedure used as the restart procedure. As the derivative of the trajectory is required for the aberration calculations, a formula is given below:

$$R'_n = \frac{1}{2h} \left\{\left(1 + \frac{h^2}{6} f_{n+1}\right) R_{n+1} - \left(1 + \frac{h^2}{6} f_{n-1}\right) R_{n-1}\right\} \tag{99}$$

Its truncation error is of the fourth order and its derivation will be found in Moses (1973) or Kasper (1982).

Determination of Aberrations by Ray Tracing. Hawkes (1980) cites a number of publications in which the authors derive the aberration coefficients by solving the exact equations of motion with sufficient accuracy to allow deduction of the aberration coefficients by matching the end-points of trajectories having different initial gradients to a polynomial. This alternative to the standard perturbation theory of aberrations has been developed further in recent years, mainly by Kasper and his team (Kasper, 1984a, 1985, 1987). This *direct method* is theoretically quite simple but can be troublesome due to its sensitivity to truncation and rounding errors. Great care must therefore be taken to avoid serious errors.

If the publications by Kasper cited above are studied carefully, it becomes clear that this direct method for determining the aberration coefficients still requires some work to perfect it as well as final verification. At least the numerical integration (also suggested by Kasper) of the exact trajectory equation (67) with the HPCD procedure for various initial values (e.g. object size γ, aperture α and additional energy $e\Delta\Phi_c$) is not always satisfactory, even if the number of trajectories which are calculated is considerably greater than the number of coefficients required. The points of intersection in the image plane are then inserted in the polynomial series expansion with respect to the initial values in the object plane (expression for image aberrations), and the coefficients of the series (aberration coefficients) are determined using a familiar least-squares fit method (LSFM). This can lead to a singular system of linear equations, which causes the calculation to be aborted unless appropriate initial conditions are selected. This disadvantage is avoided with a favorable LSFM, the Fourier analysis method, proposed by Kasper (1985). In this procedure, the lateral aberrations are expanded in terms of the azimuth of the complex aperture angle α. Moreover, Kasper uses the law, mentioned sometime previously, which states that it is possible to obtain the lateral aberration from the perturbation eikonal by partial differentiation with respect to α. In this way, he decouples the LSFM equation system to give non-singular subsystems of maximum rank 3, whose solution is not critical.

However, this does not appear to be entirely satisfactory, because the sensitivity to rounding errors, which implies that very small aberrations cannot be analyzed accurately, still remains. Kasper also suggests two possible solutions. Following his first suggestion, Kasper (1984a, 1985) does not solve the ray equation in the form given in Eq. (67) but in an incremental form directly yielding the shift $\Delta\mathbf{r}(\sigma)$ relative to another trajectory $\mathbf{r}(\sigma)$. Kasper (1987) concludes that it is always necessary to separate the aberration terms from the paraxial coordinates analytically prior to any numerical evaluation. However, he ends up with a trajectory equation which is identical to Eq. (70a) if the quadrupole terms in Kasper's trajectory equation are made zero. These terms are not taken into account in Eq. (70a). As both Chu and Munro (1982a) and Plies (1982a) have taken this equation as the starting point for the perturbation method, we must conclude that the perturbation and direct methods meet in the middle. This is a happy state of affairs.

The differential algebraic approach of Berz (1987, 1988, 1990) for the design and analysis of particle optical systems and accelerators also deserves special attention. According to Berz, this method can be used to compute transfer maps of any order for any arrangements of electromagnetic fields. As far as accuracy is concerned, it is only limited by the discretization error of the numerical field calculation. On the occasion of The 3rd International Conference on Charged Particle Optics (Toulouse, April 1989) this method was seen as a highlight of the conference and it was thought that it would give considerable impetus to modern particle optics. However, a proper comparison cannot at present be made with other methods.

3.5 Optimization

In this section, different approaches to the optimization of electron lenses and compound scanning systems (lenses plus deflectors) will be reviewed. Using Szilagyi's terminology (1988), we will classify these methods as either analytical or synthetic. Optimization by analysis means assuming that a lens or deflection system possesses certain geometrical dimensions and electric and magnetic parameters (e.g. magnetic lens excitation or the ratio of the deflection voltages of a two-stage electrical deflection system) and then calculating or analyzing its electron optical characteristics (cardinal elements, aberrations and merit functions, e.g. probe diameter). During the calculation within a superposed optimization routine, the geometric, electric and magnetic parameters (e.g. pole piece gap and excitation of a magnetic lens, axial angle or azimuth of a deflection element) are slightly modified. The new system is again analyzed, and, by iterating in this way, a system with optimal characteristics is found.

Optimization by synthesis is based on the fact that first-order properties and aberrations are completely determined by some axial functions. Therefore, instead of analyzing a vast amount of different electrode, coil and pole piece configurations, one can try to find the axial distribution that would produce the required electron optical properties and performance. This is a complicated problem. Moreover, after determining the axial functions, e.g. $\Phi(z)$, $B(z)$, the associated electrodes and pole pieces still remain to be found. This is not entirely straightforward, as small changes in the axial potential $\Phi(z)$ will cause severe changes in the contours of the electrodes. On the other hand, optimization by analysis is very tedious and may also find extremes of no interest if no reasonable guess for the design is already available. However, analysis has the advantage of always yielding realistic results.

At electron optical conferences, the author has often witnessed hard-fought disputes between the proponents of both the methods mentioned above. This is much ado about nothing as these two methods would complement each other extremely well. The first step would use synthesis, and the requirements and constraints would be changed until an acceptable axial function, e.g. $\Phi(z)$ is found. This could then be used to find the electrode shape for the lenses. The electrodes should then be reworked on the basis of aspects such as machinable design (avoiding exotic shapes) and prevention of flashovers ($|\mathbf{E}| < 10\ \mathrm{kV\,mm^{-1}}$). This design could then be used in the next step as the starting point for analytic optimization.

Reviews on electron optical optimization can be found in Hawkes (1973), Mulvey and Wallington (1973), Hawkes (1980), Szilagyi (1988), and Hawkes and Kasper (1989a). We can therefore be brief on several points, but will nevertheless deal with the latest work by Preikszas (1990) and Munack (1991) as well as with recent other work, all of which is not included in the textbooks by Szilagyi (1988) and Hawkes and Kasper (1989a).

Method of Synthesis

We will make a finer classification of the method of synthesis, on one hand using the method of calculus of variations and on the other dynamic or non-linear programming. The use of the calculus of variations can be thought of as the oldest electron optical optimization method (Tretner, 1954) in use, whereas dynamic programming was only used later (Szilagyi, 1977) to minimize electron optical aberration integrals.

The Calculus of Variations. The calculus of variations is a natural approach to the problem of synthesis, since the perturbation method of aberrations supplies the aberration coefficients as definite integrals which contain the axial field functions. It is also well suited for seeking the ultimate limits of the aberration coefficients. In the case of the important axial aberrations, these minimal possible values (with respect to the object plane) are given by the Tretner inequalities (Tretner, 1954, 1956, 1959):

Purely electrostatic lens ($B = 0$):

$$\frac{K_{c\alpha}}{L_4} \geqslant 1 \quad \text{and} \quad \frac{K_{\alpha\alpha\bar{\alpha}}}{L_4} \geqslant 1 \quad \text{for } \Phi \geqslant \Phi_{\mathrm{G}} \tag{100a}$$

$$\left.\begin{aligned} &\frac{K_{c\alpha}}{L_4} \geqslant f\left(\frac{\Phi_{\max}}{\Phi_G}\right) + V^{-2} f\left(\frac{\Phi_{\max}}{\Phi_B}\right) \\ &\frac{K_{\alpha\alpha\bar{\alpha}}}{L_4} \geqslant 0.23\left(1 + \frac{V^{-4}\Phi_G}{\Phi_B}\right) \end{aligned}\right\} \tag{100b}$$

with

$$\left.\begin{aligned} &L_4 = \frac{\Phi_G}{|\Phi'|_{\max}} \\ &f\left(\frac{\Phi_{\max}}{\Phi_i}\right) \approx \frac{2.8}{(\Phi_{\max}/\Phi_i)^{1/6}} \quad \text{for } \frac{\Phi_{\max}}{\Phi_i} < 100 \quad \text{and} \quad i = G, B \\ &f\left(\frac{\Phi_{\max}}{\Phi_i}\right) \approx 1 + \frac{7.26}{(\Phi_{\max}/\Phi_i)^{1/2}} - \frac{11.9}{(\Phi_{\max}/\Phi_i)^{3/4}} \\ &\quad \text{for } \frac{\Phi_{\max}}{\Phi_i} > 100 \text{ and } i = G, B \end{aligned}\right\} \tag{100c}$$

Purely magnetic lens ($\Phi' = 0$):

$$\left.\begin{aligned} &\frac{K_{c\alpha}}{L_1} \geqslant \frac{\pi}{4}\left(1 + V^{-2}\right) \\ &\frac{K_{\alpha\alpha\bar{\alpha}}}{L_1} \geqslant 0.33\left(1 + V^{-4}\right) \end{aligned}\right\} \tag{101a}$$

$$\frac{K_{c\alpha}}{L_2} \geqslant 0.5 \qquad \frac{K_{\alpha\alpha\bar{\alpha}}}{L_2} \geqslant 0.24 \qquad \frac{f_G}{L_2} \geqslant 0.8 \quad \text{for } V = \infty \tag{101b}$$

$$\frac{K_{\alpha\alpha\bar{\alpha}}}{L_3} \geqslant 0.6 \quad \text{if } B(z_G) = 0 \tag{101c}$$

where

$$\left.\begin{aligned} &\frac{1}{L_1} = B_{\max}\sqrt{\frac{e}{8m\Phi}} \approx \frac{0.15}{\text{cm}} \frac{B_{\max}}{\text{Gauss}} \sqrt{\frac{V}{\Phi}} \\ &\frac{1}{L_2} = \left|\frac{B'}{B}\right|_{\max} \qquad \frac{1}{L_3} = \left|B'_{\max}\right|^{1/2} \left(\frac{e}{8m\Phi}\right)^{1/4} \end{aligned}\right\} \tag{101d}$$

Apart from these inequalities, Tretner (1959) also provides helpful plots of axial aberration coefficients against focal length, in which the regions of existence for electrostatic or magnetic lenses are shown. This work also contains the axial functions $\Phi(z)$ and $B(z)$ for optimal electron lenses of pure electrostatic and pure magnetic type. Mulvey and Wallington (1973) have mentioned that no lens has ever been found outside the Tretner limits and so, in spite of a number of controversies, Tretner's theory can be assumed to be correct. On the other hand, Mulvey and Wallington have found a range of lenses which do not satisfy the optimal Tretner field curve. However, their aberration coefficients do not differ greatly from the minimal values given by Tretner. This means that the optimum appears to be very flat and that it could be found by conventional trial and error methods instead of involved techniques of mathematical optimization. However, the situation looks rather different when further constraints, e.g. that lenses should be coma-free, are introduced or superposed electrostatic–magnetic lens fields are considered. The latter are the object of great interest in the field of low-voltage electron optics.

The publications of Moses (1973), Rose and Moses (1973), and Preikszas (1990) are further sources dealing with the calculus of variations. Moses's advanced procedure (1973) searches the optimal $B(z)$ curve for minimal spherical aberration. The constraints are fixed object and image positions, keeping the axial field strength $B(z)$ below a reasonable upper bound, and a vanishing coefficient of anisotropic coma (imaginary part of the complex coma coefficient). Together with the paraxial equation, all conditions are combined in one variational principle and the corresponding Euler equations are solved numerically by Numerov's method. Moses also gives, *inter alia*, the pole pieces for a probe lens with minimum spherical aberration and field-free object location. This lens differs from a conventional pinhole lens by having tapered pole pieces inside the lens. (Moses's lens is not the only one that shows the fallacy of the occasionally expressed opinion that no practicable lenses have been designed by using the synthesis method.)

Preikszas (1990) was able to show that with combined electrostatic–magnetic lenses the Euler–Lagrange formalism leads to a very complicated system of differential equations when only the spherical aberration is minimized. This and the boundary conditions and constraints mean that the problem is extremely involved. He therefore developed a new *direct* method which allows the combined electrostatic–magnetic field, i.e. $\Phi(z)$ and $B(z)$, to be expressed by a fixed number of sample points and makes it possible to optimize the sum of the squares of the spherical aberration, axial chromatic aberration, coma and diffraction (rule of thumb for the minimum obtainable off-axial spot size). By varying the sample points point-by-point and by observing a number of constraints, the optimal solution is obtained after iterating for some time. Preikszas obtained good agreement when he compared his direct method and the classical calculus of variations (Moses, 1973) in the case of purely magnetic fields. The advantage of the direct method of Preikszas is its simple handling of constraints; a disadvantage is the long computation time, but this is not crucial in view of the power of modern computer systems.

Some of the axial functions $\Phi(z)$ and $B(z)$ found by Preikszas cannot be converted into practicable electrodes and pole pieces, but others can. Preikszas has also tried to decide whether a purely electrostatic immersion lens, as suggested by Zach (1989), or an electrostatic immersion lens with a superposed magnetic lens, as suggested by Frosien et al. (1989), can further increase resolution in the field of low-voltage electron optics. In both cases, he selected a lens field length of 30 mm, an acceleration potential at the probe of 500 V, an immersion ratio of 10 and a (half) PE energy spread of 0.2 eV. Moreover, he specified a field-free probe separation of 2.5 mm and only permitted a maximum axial electric field strength of 2 kV mm^{-1} and an axial magnetic flux density of 0.2 T. The axial functions obtained in both cases $\Phi(z)$, $\Phi'(z) = -E(z)$, $B(z)$ and the fundamental trajectory $u_\alpha = r_\alpha$ are shown in Fig. 19. It should be noted that the ordinate scales for $E(z)$ are different in Fig. 19A and Fig. 19B. With the magnetic field (Fig. 19B) the minimum spot radius only improves by 3.4% compared with the purely electrostatic lens field of Fig. 19A, because there was only a slight improvement in the axial chromatic aberration, which is crucial in this case. If the comparison was made for the same focal length and not the same field length, the combined electrostatic–magnetic system would come off somewhat better. The advantage, however, lies in the considerably lower electrostatic field strength (factor 8) and the important ability of a system of this kind to operate at higher acceleration voltages because of the presence of the magnetic lens. This is essential for, say, EDX (energy dispersive X-ray analysis) or EBIC (e-beam induced current analysis). For inspecting semiconductor probes, it would be best to provide a low-voltage SEM with both the methods of analysis mentioned above. As far as an electron optical defect review is concerned, the defect must be analyzed chemically as well as geometrically, and EDX is well suited for this. In practice, the condition that the target plane be field-free can be dropped for the magnetic field but not the electrostatic field. One of the consequences of this is an increase in the resolution for the combined electrostatic magnetic case. The magnetic field at the probe location would first and foremost provide a better through-lens detection of the SEs with the probe tilted.

Dynamic Programming and Non-linear Programming. A very different technique has been explored by Szilagyi (1977), who combined the search routines developed for dynamic programming (see Bellman, 1957) with a very sim-

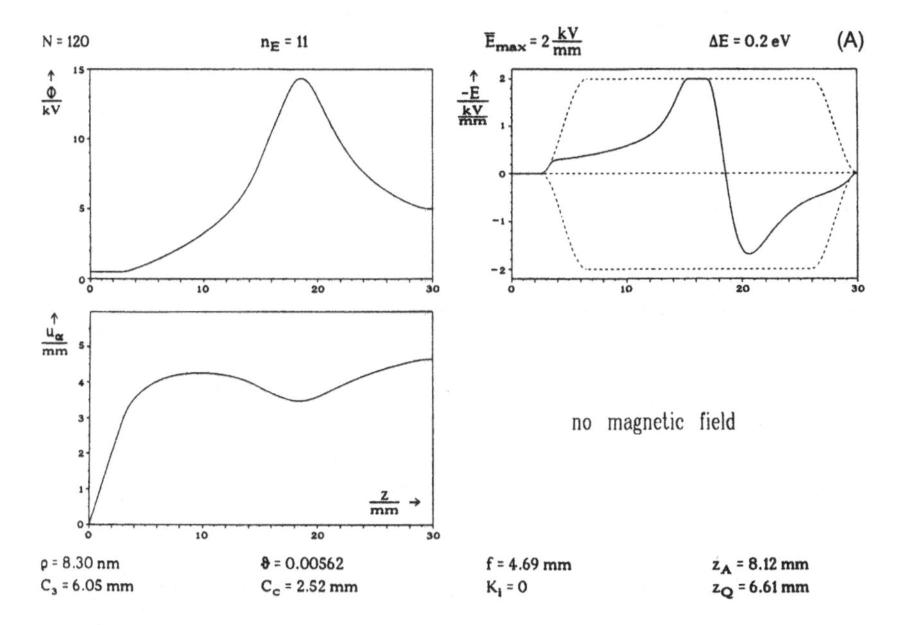

Figure 19 Optimal axial distributions $\Phi(z), E(z) = -\Phi'(z), B(z)$ and fundamental ray $u_\alpha(z) = r_\alpha(z)$ for minimal spot radius ρ in the specimen plane ($z = z_G = 0$): (A) without magnetic field, (B) with magnetic field. The additional numerical data shown have the following meanings or are linked as follows to the quantities introduced above: $N + 1$, number of sample points used; n_E, n_B, number of sample points close to the object plane ($z = z_G = 0$) which are free of the electrostatic or magnetic field, respectively; E_{max}, B_{max}, maximum permitted field strengths (constraints); $\Delta E = e\Delta\Phi_c$; $\vartheta = |\alpha|$; $f = f_G$; $C_3 = K_{\alpha\alpha\bar{\alpha}}$; $C_c = K_{c\alpha}$; $K_i = |Jm(K_{\alpha\alpha\bar{\gamma}})|$; z_A, aperture plane (real intersect of the r_γ-trajectory with the optical axis) for $\mathrm{Re}(K_{\alpha\alpha\bar{\gamma}}) = 0$; z_Q, virtual intersect with the optical axis associated with z_A. *From Preikszas (1990).*

ple electrostatic or magnetic lens model. Szilagyi has improved this method over the years and used it for practical applications. A full description is given in his textbook (Szilagyi, 1988) and will be useful to any reader who wishes to go into the matter more deeply.

Szilagyi, Yakowitz, and Duff (1984) proposed another procedure for optimizing the spherical aberration coefficient of electrostatic lenses. They divided the axial length of the lens into different regions and made the assumption that the axial potential can be represented in each region by a simple polynomial expression. The requirement that $\Phi(z)$ and its lower derivatives be continuous leads to some relationships between the coefficients of the polynomials at neighboring regions. Szilagyi et al. (1984) define the continuity requirements in such a way that the set of coefficients at the highest-degree terms remain free. The problem is then reduced

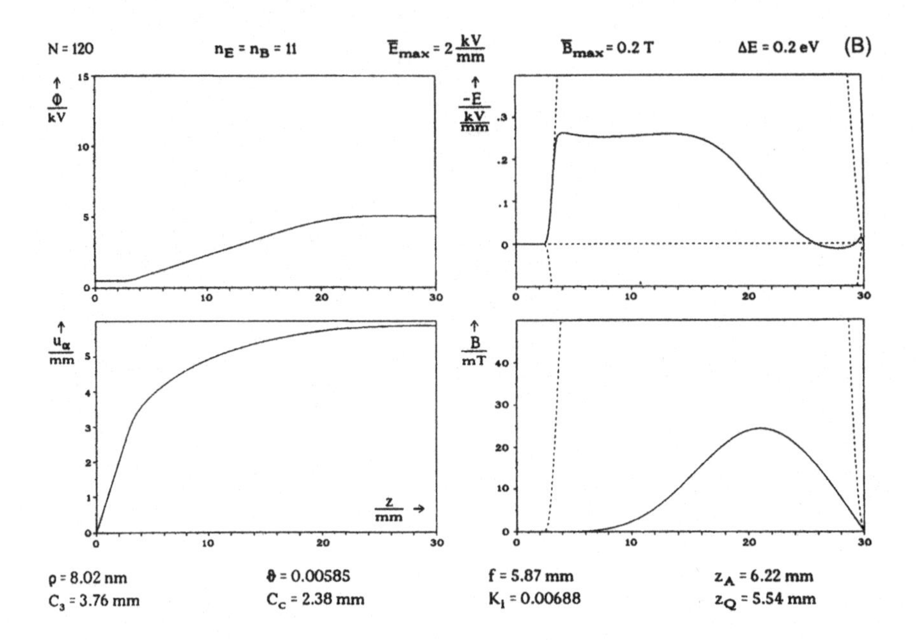

Figure 19 *(continued)*.

to finding coefficients satisfying the paraxial ray equation and the constraints, and then minimizing the aberration integral. The minimization can be performed by applying any non-linear programming technique, e.g. a quasi-Newton algorithm. For non-linear programming, i.e. non-linear optimization, see, for example, Künzi, Tzschach, and Zehnder (1967), Künzi, Krelle, and von Randow (1979), Bialy and Olbrich (1975), Gill, Murray, and Wright (1981), Dennis (1984), and Rao (1984). By utilizing methods of non-linear programming, one can expect only local convergence to a local minimum.

Scheinfein and Galantai (1986) have generalized the method of Szilagyi et al. (1984) (more appropriate figure of merit, e.g. electron probe diameter) and make some observations on the choice of minimization algorithms and penalization procedures. Szilagyi (1988) discusses the remaining problem of reconstructing the electrodes and pole pieces from the optimized axial field curves. He also describes a reconstructed electrostatic triple-electrode lens. Further examples illustrating reconstructed electrostatic lenses can be found in Scheinfein and Galantai (1986).

Glatzel and Lenz (1988) have optimized electrostatic immersion lenses by adapting a method which has been successfully applied in light optics

(Glatzel, 1961). We will not go into these local methods here, but merely mention that Glatzel and Lenz (1988) have found potential distributions $\Phi(z)$ giving coefficients of spherical aberration which are less by a factor of two than those obtained by Szilagyi (1983) with the global method of dynamic programming (and the same constraints). Moreover, the values obtained by Glatzel and Lenz (1988) are very close to the ultimate limit. As they did not describe the axial potential distribution using cubic splines, but instead used four or five circular aperture potentials (Lenz, 1956b), the reconstruction of real electrodes is considerably easier and more precise.

Method of Analysis

The existing analytic optimization methods for electron optics can be classified according to the problem (lenses, compound scanning systems), the figure of merit, the constraints and the selected linear or non-linear optimization algorithm. In the case of linear optimization (= linear programming), the simplex algorithm is a general method of solution. Unfortunately, the electron optical optimization problem is not in general linear with respect to the figure of merit or the constraints. There is no general method of solution for non-linear optimization (= non-linear programming). Therefore, a suitable non-linear programming procedure must be selected for each separate case, unless one wishes to embark on the tedious process of piecewise linearization.

Optimization of Lenses. Gu and Shan (1984) used the complex method for constrained optimization (Richardson & Kuester, 1973) to minimize the spherical aberration of an extended electrostatic field lens. Adriaanse, van der Steen, and Barth (1989), and van der Steen and Barth (1989) developed a practical method of optimizing electrostatic lenses which they used to substantially improve some existing designs. An important property of their lens model is that the electrodes generate only lower-order derivatives which are a solution to the Laplace equation up to the second order. For this reason, they have named this method the second-order electrode method (SOEM). An application-dependent figure of merit is used and is a function of the spherical and axial chromatic aberration coefficients. The minimization is performed using a variable metric optimization routine.

Optimization of Compound Scanning Systems. More aberration coefficients are found in compound scanning systems than in round lenses. The geometric and electrical parameters of the deflection elements and the position of the deflection elements with respect to the round lens are additional variation parameters. The dependence of the aberrations on these parameters is highly non-linear, so that algorithms for non-linear optimization are used. This means, of course, that only a local minimum, which is often strongly dependent on the initial parameters, can be found. Munro (1975) used the method of Powell (1965) for minimizing the sum of squares of functions of several variables. Later, Chu and Munro (1981, 1982b) employed the damped least-squares method (Levenberg, 1944) for optimizing compound scanning systems for electron lithography. This method, which has been successfully used previously in light optical design, seems to be one of the most effective optimization algorithms available at present. Full details of this method can be found in the paper by Chu and Munro (1982b).

Finally, we want to note two other optimization methods. The first method originates from a new field of science, bionics, and uses evolution strategies for the numerical optimization of technical systems (Rechenberg, 1973; Schwefel, 1977). This mutation selection method has been successfully used to solve optimization problems which largely come from the field of fluid mechanics, e.g. two-phase supersonic jet engines. However, the mutation selection method only converges satisfactorily if the mutation step width is chosen correctly. Investigations have shown that this step width is optimal when, on average, every fifth mutation is successful. This can be used to derive a method of step width control. It may perhaps be rewarding to use this method for optimizing multiparameter electron optical systems. The second method, which also deserves our attention, is the global optimization of electron optical systems by Munack (1991) with the aid of interval arithmetic (Moore, 1966; Alefeld & Herzberger, 1974). Unlike dynamic programming, this method is not limited by discretization.

4. COULOMB INTERACTION AND ELECTROSTATIC CHARGING

Coulomb interaction in the PE beam and electrostatic charging of an insulating target are two significantly limiting influences in e-beam testing and electron optical inspection. Coulomb interaction, particularly in the case of high beam current and low beam voltage, causes a deterioration in the PE optical performance data such as brightness, resolution or probe diameter.

If the specimen becomes electrostatically charged, no significant image of the specimen can be generated and erroneous measurements of the voltage states are also obtained in e-beam testing. A material-dependent primary beam energy of between 500 eV and 3 keV is required for the charge-free investigation of insulating materials. It is then usually possible to ensure that one SE is emitted by the specimen for every PE (second neutral point of the SE emission). This requirement for a low primary beam voltage gives rise inevitably to a deterioration of the electron optical resolution. This is first because the chromatic aberration of lenses and deflectors is inversely proportional to the primary beam voltage, and secondly because the diverse harmful effects of the Coulomb interaction of the primary beam electrons are also greater at low acceleration voltage. The requirement of short measurement or inspection times can only be fulfilled by means of a higher probe current. Thus, with increasing beam current and electron densities, Coulomb interaction becomes the dominant factor limiting the performance of a low-voltage electron optical column.

4.1 Boersch Effect

One consequence of the Coulomb interactions is the anomalous energy broadening of the electrons in the electron optical column, known as the Boersch effect (Boersch, 1954). This *(energetic) Boersch effect* leads to a deterioration in the resolution of the column due to chromatic aberrations from lenses and deflectors. Moreover, as a result of stochastic Coulomb interactions a direct anomalous broadening of the probe diameter also occurs. This phenomenon is often known as the spatial Boersch effect. In the anomalous spatial effect, a distinction can be made between the *trajectory displacement effect* and *statistical angular deflections* (angular spread) (Loeffler & Hudgin, 1970; Jansen, 1988a). Both these terms will be used in the following treatment.

All the effects considered here result from residual stochastic Coulomb interactions in the e-beam after subtraction of the mean Coulomb interaction caused by the (homogeneous) space charge distributions. For more details on the space charge optics and the defocussing and spherical aberration caused by the space charge, reference should be made to the literature (e.g. Wendt, 1948; Pierce, 1949; Glaser, 1952; Hutter, 1967; Nagy & Szilagyi, 1974; Lawson, 1977; Jansen, 1988a).

The various phenomena associated with the Coulomb interaction in e-beams have been extensively investigated in the last 35 years, both experimentally and theoretically, and this has also led to a certain amount of controversy. For the numerous references in the literature and historical notes, reference should be made to the review article by Rose and Spehr (1983) and the dissertation by Jansen (1988a), which has meanwhile been published by Academic Press (Jansen, 1990). At the end of the section on the energetic and spatial Boersch effects in their textbook on electron optics, Hawkes and Kasper (1989b) note that, despite the importance of these effects and the many experimental and theoretical investigations that have been devoted to them, our present knowledge about their origin and behavior is still unsatisfactory. This statement may well still be true today, but thanks mainly to the work of Jansen (1988a), which had not yet been taken into account by Hawkes and Kasper, major strides have now been made in dealing with this problem. For the programs developed by Jansen, *MONTEC* (Monte Carlo simulation) and *INTERAC* (analytical model), marketed by the Delft Particle Optics Foundation, provide the designer of particle optical systems with two tools for the simulation of the Coulomb interaction in low- and medium-density non-relativistic beams in a drift space. With these programs, Jansen (1988b) successfully simulated the (energetic) Boersch effect and the trajectory displacement effect between the first beam-shaping aperture and the target plane in the electron optical columns of the Perkin–Elmer Aeble 150 and IBM EL3 electron lithography devices. Nevertheless, neither a complete theory nor a simulation program exists for calculating the Coulomb interaction in electron guns. However, the special case of the Coulomb interaction in an e-beam which is accelerated in a uniform axial electrostatic field has already been dealt with by Knauer (1981), Sasaki (1986), Munro (1987b), and Jansen (1988a).

When calculating the Coulomb interaction, a distinction in principle must be made between analytical models and the numerical simulation based on the Monte Carlo method. All Monte Carlo calculations have the drawback of not yielding any insight into the physical processes involved. Attention should therefore be drawn to the work of Loeffler and Hudgin (1970), El-Kareh and Smither (1979), Groves, Hammond, and Kuo (1979), Sasaki (1979, 1982), Munro (1987b), and Jansen (1988a), in which the Monte Carlo method is employed. Due to the complexity of the stochastic many-body problem, all analytical approaches involve some degree of approximation. A number of theories have been developed which differ widely in the approximations used. Comparisons between the different analytical models are to be found in Rose and Spehr (1983) and in Jansen (1988a, 1990). Only two models will be mentioned here which seem to be the best approximations known:

- The *closest encounter* or *nearest-neighbor approach* introduced by Rose and Spehr (1980). They assumed that the main contribution stems from the collision between the reference particle and the particle which has the smallest impact parameter with respect to it.
- The *extended two-particle model* or *multiple independent collision approach* of van Leeuwen and Jansen (1983). They break down the n-particle collision problem into a succession of n independent two-particle interactions.

For many applications, both models provide results in reasonable agreement. The validity range of the extended two-particle model, however, appears to be somewhat greater (e.g. see Jansen, 1988a). The most comprehensive collection of comparatively simple formulae is to be found in Jansen (1988a). In some formulae, however, the constants have been determined by parameter fitting of numerical data. Some useful formulae from Jansen on the (energetic) Boersch effect in beam segments with a crossover will be reproduced here because the electron density and thus the Coulomb interaction is strongest in the crossovers of the beam path. In this section and the next, Jansen's nomenclature will be employed. Before we move on to the actual formulae for the energy broadening, the beam parameters which they contain will be indicated:

Scaled linear particle density $\bar{\lambda}$:

$$\bar{\lambda} = \frac{m^{1/2}}{2^{7/2}\pi\varepsilon_0 e^{1/2}} \frac{I}{\alpha_0^2 V^{3/2}} \tag{102a}$$

Scaled crossover radius $\bar{r}_c$:

$$\bar{r}_c = \frac{8\pi\varepsilon_0}{e}\alpha_0^2 V r_c \tag{102b}$$

Crossover position parameter S_c:

$$S_c = L_1/L \tag{102c}$$

Characteristic beam geometry quantities K, K_1, and K_2:

$$K = \alpha_0 L/(2r_c) \qquad K_1 = 2KS_c \qquad K_2 = 2K(1 - S_c) \tag{102d}$$

In these equations, m is the electron mass, e the elementary charge, ε_0 the electric constant, V the beam voltage, and I the beam current. The

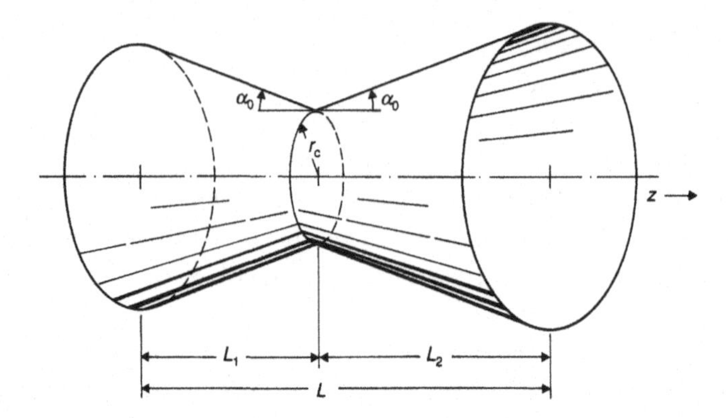

Figure 20 Definition of the experimental parameters for a beam segment with a crossover in drift space. The beam dimensions are specified by the crossover radius r_c, the beam semi-angle α_0, and the length L of the beam segment, which can be divided into two parts before and after the crossover, of lengths L_1 and L_2. *Redrawn from Jansen (1988a).*

meaning of the geometric parameters α_0, r_c, L_1, and L of a beam segment with a crossover in drift space can be seen in Fig. 20.

Let us first consider the important case of a narrow crossover, i.e. $K \gg 1$, located somewhere near the middle of the beam segment ($S_c \approx 0.5$) so that $K_1, K_2 \gg 1$ is also true. Let us assume that the scaled linear particle density is high, $\bar{\lambda} \gg 10^{-2}$, that we shall have strong collisions which will cause a spread in axial velocities with a *Gaussian distribution*. According to Jansen (1988a), the following will then be true for the relative energy spread if the full width at half maximum (FWHM) is taken:

$$\frac{\Delta E_{\text{FWHM}}}{E} = \left(\frac{8(\ln 2)^2 m}{\varepsilon_0^2 e}\right)^{1/4} \sqrt{\frac{I}{V^{3/2}}} \quad \text{for } \bar{r}_c \leqslant 10 \tag{103a}$$

$$\frac{\Delta E_{\text{FWHM}}}{E} = \left(\frac{8(\ln 2)^2 m}{\varepsilon_0^2 e}\right)^{1/4} \left\{1 + \frac{\pi \bar{r}_c}{[2\ln(0.8673(114.6 + \bar{r}_c))]^2}\right\}^{-1/2} \sqrt{\frac{I}{V^{3/2}}}$$
$$\text{for any } \bar{r}_c \tag{103b}$$

$$\frac{\Delta E_{\text{FWHM}}}{E} = 1.450 \frac{(me)^{1/4}}{\varepsilon_0} \left\{1 + 0.217 \ln\left(1 + \frac{\bar{r}_c}{114.6}\right)\right\} \frac{I^{1/2}}{r_c^{1/2} \alpha_0 V^{5/4}}$$
$$\text{for } \bar{r}_c \geqslant 100 \tag{103c}$$

If, in the case of a narrow crossover, the particle density assumes intermediate values with $\bar{\lambda} \ll 10^{-2}$, then the weak complete collisions are dominant and the result is a *Lorentz distribution* with

$$\frac{\Delta E_{\mathrm{FWHM}}}{E} = \left(\frac{2m}{\varepsilon_0^2 e}\right)^{1/2} \frac{I}{\alpha_0 V^{3/2}} \tag{104}$$

The case of a beam segment with a crossover of arbitrary dimensions is even more difficult to describe analytically. For intermediate values of $\overline{\lambda}$ and small K values, however, we find the following relatively simple formula in Jansen (1988a):

$$\left.\begin{aligned} &\frac{\Delta E_{\mathrm{FWHM}}}{E} = 1.323\frac{m^{1/3}}{\varepsilon_0} H(K) \frac{I^{2/3}}{r_c^{1/3}\alpha_0 V^{4/3}} \\ &\text{with} \\ &H(K) = \left\{1 - \frac{1}{2(1+K_1)^{1/3}} - \frac{1}{2(1+K_2)^{1/3}}\right\} \end{aligned}\right\} \tag{105}$$

In this special case, weak incomplete collisions are dominant, resulting in a *Holtsmark distribution*.

For small $\overline{\lambda}$ values with $\overline{\lambda} \ll 1/(2K\overline{r}_c)$, a beam with a crossover is referred to as a *pencil beam*. For the relative energy broadening, Jansen (1988a) then gives the simple formula

$$\frac{\Delta E_{\mathrm{FWHM}}}{E} = 0.11 \frac{m}{\varepsilon_0 e^2} \frac{LI^2}{V^2} \tag{106}$$

Notice that this equation is independent of the crossover radius r_c and the aperture α_0.

For the cases not dealt with here with other combinations of beam parameters $\overline{\lambda}$, $\overline{r}_c$, and K, reference should be made to the work of Jansen (1988a, 1990), as in those cases the analytical formulae are even more complex than in the ones dealt with here. Nevertheless, the simple formulae reproduced here from Jansen (1988a, 1990) are applicable for many crossover ranges in e-beam testing and electron optical inspection. Furthermore, the equations reveal important dependencies of experimental parameters such as I and V which provide indications for reducing the Boersch effect. We shall return to these later.

4.2 Trajectory Displacement Effect

According to Jansen (1988a), the calculation of statistical angular deflections (angular broadening) is primarily of interest for a better physical understanding of trajectory displacement effects. As mentioned by

Weidenhausen, Spehr, and Rose (1985), the stochastic ray deflections can limit the edge resolution of microlithographic systems using the variable-shaped beam method and diminish the contrast in high-resolution transmission electron microscopy. This, however, is beyond the scope of the present work, so we shall not consider the angular spread resulting from the Coulomb interaction of beam electrons.

We also consider only the (special) trajectory displacement effect that occurs in a homocentric beam segment with a crossover, i.e. $r_c = 0$ or $K = \infty$. In the case of a beam segment with a crossover of arbitrary dimensions, there is no change in the dependence of the experimental parameters L, α_0, V, and I. But the prefactor of the corresponding analytical equations, in which further parameters such as S_c and K are contained, is more complicated. The following relatively simple equations for the case of a punctiform crossover ($r_c = 0$) have been taken from the work of Jansen (1988a). First, however, we must introduce two more parameters in addition to those defined in Eqs. (102a)–(102d). The image plane position parameter S_i is defined as $S_i = L_i/L$, where L_i is the distance from the entrance plane (of the beam segment) to a plane that is optically conjugated to the target plane. This plane, referred to as the image plane, can be located either inside or outside the beam segment. In source (crossover) imaging systems, the image plane coincides with the crossover plane, but this is not true for all beam segments if Koehler illumination is used. The pencil beam factor χ_c is related to the already familiar parameters $K, \bar{r}_c$, and $\bar{\lambda}$ by the relation $\chi_c = 2K\bar{r}_c\bar{\lambda}$.

For $\bar{\lambda} \gg 10^{-3}$ ($\bar{\lambda} \gtrsim 0.05$) and $\chi_c \gg 1$, we obtain a two-dimensional *Gaussian distribution* for the lateral trajectory displacement Δr due to the dominant strong complete collisions. For $S_c = S_i$, its half-value width can be approximated by means of the following equation:

$$\Delta r_{\mathrm{FWHM}} = 0.158 \frac{e^{1/12} m^{1/4}}{\varepsilon_0^{5/6}} \frac{[1 + 0.682(S_c - 0.5) - 0.739(S_c - 0.5)^2]^{1/2}}{(1 + 1.4/v_0^{*8/7})^{7/4}} \times \frac{I^{1/2} L^{2/3}}{\alpha_0 V^{13/12}} \tag{107}$$

with the parameter $v_0^* = (2\pi\varepsilon_0/e)^{1/3} \alpha_0 V^{1/3} L^{1/3}$.

The regime of *weak complete collisions* becomes manifest only for $S_c \neq S_i$. This case is unimportant for e-beam testing and electron optical inspection because in these cases the crossover is imaged, i.e. $S_c = S_i$ is true. For $\bar{\lambda} \lesssim 0.01$ and $\chi_c \gg 1$, the weak incomplete collisions are dominant, resulting in a *Holtsmark distribution*. For the half-value width of the trajectory displacement, according to Jansen (1988a), with $S_c = S_i$ (and $K = \infty$ as generally agreed above), a good approximation is

$$\Delta r_{\mathrm{FWHM}} = 0.137 \frac{m^{1/3}}{\varepsilon_0} |S_c^{2/3} + (1 - S_c)^{2/3}| \frac{I^{2/3} L^{2/3}}{\alpha_0^{4/3} V^{4/3}} \tag{108}$$

Finally, for a pencil beam with $\chi_c \ll 1$, we have

$$\Delta r_{\mathrm{FWHM}} = 6.92 \times 10^{-5} \frac{m^{3/2}}{\varepsilon_0 e^{7/2}} |4 - 6(S_c + S_i) + 12 S_c S_i| \frac{\alpha_0 I^3 L^3}{V^{5/2}} \tag{109}$$

Let us leave it at these formulae and refer again to the work of Jansen (1988a) for the beam segment with a finitely large crossover ($r_c \neq 0$). Reference should also be made to the more recent work of Weidenhausen and Spehr (1989) on angular spread and of Berger, Spehr, and Rose (1990) on the trajectory displacement effect. The results of Berger et al. (1990) are in close agreement with the theory of Jansen (1988a). In the formulae given above, we have specified the FWHM and not the full width median or the rms (root mean square) width. This appears to be most useful for physical reasons, as the rms width, for example, is dominated by the tails of the displacement distribution. It thus supplies a correct width measurement only in the case of a Gaussian distribution. Conversion calculations between the three width measurements are to be found in Jansen (1988a).

4.3 Means of Reducing the Boersch Effect and the Trajectory Displacement Effect

The Boersch effect in an electron optical column is determined mainly by the Coulomb interaction of the electrons in the individual beam segments with a crossover. If a spread in axial velocities with a Gaussian distribution is generated by the Coulomb interactions in all individual beam segments, then, according to Jansen (1988a), the individual FWHMs must be summed quadratically; for the summation in the case of other distributions, see Jansen (1988a). To minimize the total Boersch effect, the column should have as few crossovers as possible. Furthermore, the beam geometry should be chosen in such a way that each individual contribution remains small. In the case of Gaussian distributions, the total FWHM is strongly dependent on the largest single FWHM, which therefore deserves our particular attention.

If, in Eq. (103c), we take into account the Helmholtz–Lagrange relation in the form $\sqrt{V} r_c \alpha_0 = \text{constant}$, and approximate the terms within the braces with 1, we then obtain

$$\Delta E_{\mathrm{FWHM}}/E \sim r_c^{1/2} I^{1/2} / V^{3/4} \tag{110}$$

This means that, for a constant beam current I and constant acceleration voltage V, the largest contribution to the Boersch effect comes from the crossover with the largest diameter $2r_c$. In demagnifying probe-shaping systems, the first crossover has the largest diameter. Thus, it is of advantage in such a demagnifying system to make this first crossover as small as possible.

Eqs. (103a)–(106) all show that the Boersch effect increases with rising beam current I and falls with rising acceleration voltage. As the probe current in most SEM-type devices is smaller by a factor of 100–1000 than the initial beam current in the gun, the beam current should be reduced almost to its final value in the upper part of the electron optical column in order to reduce the Boersch effect. The dependence of the Boersch effect on the acceleration voltage makes it advisable in low-voltage electron optical columns to keep the electron energy high (e.g. 5–10 keV) as long as possible and only to decelerate the electrons to the desired final energy (e.g. 0.5–1 keV) in the final lens, an immersion-type objective lens.

Even if the above equations have no quantitative validity for the Boersch effect in the electron gun, it is still possible to infer from them that a crossover *within* the gun supplies the largest contribution to the Boersch effect. In the first instance, the beam current is still highest here, the electron energy is still low before the anode and, in the case of a demagnifying system, this crossover has the greatest diameter. One step for reducing the Boersch effect particularly worthy of note here is the use of an electron gun with a virtual crossover. An additional step already mentioned by Rose and Spehr (1983) is to locate the first (real) crossover of the column on the far side of the beam-limiting aperture. In this case, the masked-out electrons make a negligible contribution to the Boersch effect.

In addition to the circular crossover, Rose and Spehr (1980, 1983) also considered an astigmatic crossover and found that the Boersch effect is weaker in this case (Fig. 21). Thus, the use of astigmatic crossovers is also a means of reducing the Boersch effect. Scherzer (1971) has already pointed out that, in a demagnifying system, it is important to make the first crossover astigmatic to significantly lower the Boersch effect. Unfortunately, astigmatic systems are difficult to adjust and usually contain more elements than rotationally symmetrical ones. Nevertheless, Nerstheimer (1979) reported that this is not necessarily the case. In the 30 AX television CRT

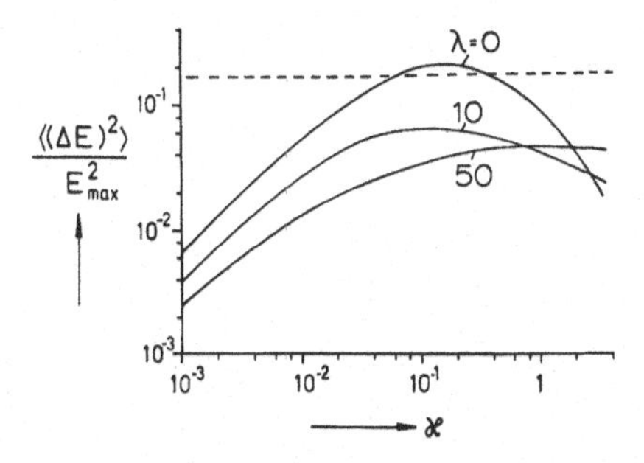

Figure 21 Mean quadratic energy broadening as a function of the current parameter $\varkappa$ for three values of the astigmatism parameter λ. A muffin-tin distribution is assumed for the transmitted current density in the aperture plane. The dashed line indicates the thermodynamic limit. *From Rose and Spehr (1980).* $E_{max} = 2\alpha_0$ eV is the maximum energy which can be transferred to each of the two mutual scattering electrons of initial energy eV. The relationships $\varkappa = \overline{\lambda}/2$ and $\lambda = 4\overline{r}_c = 32\pi\varepsilon_0\alpha_0^2 r_c V/e$ exist between the parameters $\varkappa$ and λ from Rose and Spehr (1980) and the parameters from Jansen (1988a) used in the text. In the case of astigmatic imaging, r_c must be replaced by $\alpha_0\Delta/2$, where Δ is the astigmatic difference (distance between the two astigmatic focal lines) and $\alpha_0\Delta$ is the diameter of the circle of least confusion (located midway between the focal lines).

system from Philips/Valvo, the shape of the Wehnelt electrode (grid G1) has been modified so that two astigmatic crossover lines occur in the area of low electron velocity instead of one stigmatic crossover. This results in a considerable reduction in Coulomb repulsion and in increased image sharpness. Rose and Spehr (1983) also mention the use of polarized electrons for a significant reduction of the Boersch effect in the case of diffraction-limited probes.

The above-mentioned methods for reducing the Boersch effect also help, at least marginally, to reduce the trajectory displacement effect. In a demagnifying probe-forming system, the trajectory displacement effect of the last beam section between the objective lens and the target is usually the greatest. This is because the trajectory displacement effects in all the previous beam sections with a crossover are demagnified. From Eqs. (107)–(109) it can be seen that the distance L, to be interpreted in this case as the working distance between the image side principal plane and the target, must be kept as small as possible. The reduction in the normalized brightness (brightness divided by acceleration potential) observed in the case of low voltage is probably one result of the trajectory displacement effect in the crossover located within the gun. (In fact, according to Eq. (60), the normalized brightness should not be dependent upon the acceleration

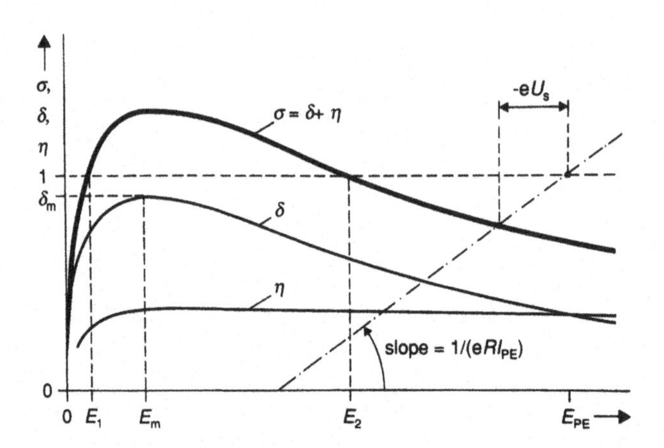

Figure 22 Dependence of the secondary electron yield δ, the back-scattering coefficient η and the total yield σ on the PE energy $E_{PE} = eU_{PE}$ (schematic). For an energy $E_{PE} > E_2$, negative charging occurs and the surface potential $U_s < 0$ can be constructed as the intersection of the total yield σ and the straight line shown, whose slope is $1/eRI_{PE}$, I_{PE} being the PE current and R the leakage resistance.

voltage.) As, according to Eqs. (107)–(109), the trajectory displacement effect of a beam segment with a crossover in field-free space increases with a decrease in acceleration voltage, this explanation appears plausible, even if these equations are not actually applicable for a crossover located within the gun. However, experiment appears to confirm this explanation. Frosien and Plies (1987) used a triode gun at a high acceleration voltage (5 kV) and decelerated the electrons to their final energy (1 keV) in an immersion condenser lens. The measured brightness was greater by a factor of 6 than that reported by Yamazaki, Kawamoto, Saburi, and Buchanan (1984) for a triode gun driven directly with an acceleration voltage of 1 kV and otherwise identical conditions (⟨100⟩-LaB_6 cathode, $T = 1500\,°C$).

4.4 Electrostatic Charging

Whether an insulating target becomes positively or negatively charged or not at all depends on the total electron emission yield $\sigma = \delta + \eta$, which is composed of the SE yield δ and the back-scattering coefficient η. Fig. 22 shows the fundamental dependence of the total yield σ on the PE energy $E_{PE} = eU_{PE}$ with normal incidence of the PEs. As can be seen, there are two PE energies E_1, E_2 for which $\sigma = 1$ and thus for which the target does not become charged. E_1 has low values of a few hundred volts and does not allow stable non-charging operation. So here we shall consider only

Table 3 Maximum SE yield δ_m, PE energy E_m at which δ_m occurs, and PE energy E_2 at which the total (SE plus back-scattered electron) yield $\sigma = 1$ of some materials used in IC fabrication and electron optics

Material	Constant		
	δ_m	E_m (keV)	E_2 (keV)
Carbon	0.94–1.0	0.3–0.55	0.65
Aluminum	0.97–1.17	0.30–0.40	1.05
Silicon	0.9–1.10	0.25–0.40	1.15
Chromium	1.0–1.16	0.48–0.60	1.8–2.0
Iron	1.1–1.3	0.40	1.27
Copper	1.1–1.3	0.50–0.75	2.74–2.8
Be–Cu bronze	2.20–5.0	0.3–0.4	
Molybdenum	1.0–1.24	0.40–0.65	2.23–3.0
Silver	1.0–1.4	0.70–0.80	3.2–5.8
Gold	1.31–1.45	0.7–0.8	7.8–8.27
Al_2O_3	2.60–3.0	0.30–0.4	
SiO_2	2.5	0.42–0.5	3.0
Glass passivation	2–3	0.3–0.42	2.0
Ni silicide	1.97	0.8	6.5
GaAs	1.20	0.6	2.6
PVC			1.65
Teflon–FEP	2.21–3.0	0.3–0.4	1.82–1.9
Kapton (polyimide)	2.10	0.15	
HPR resist	1.09	0.37	0.55
PBS resist			0.70
AZ1470 resist			0.9–1.10

the influence of the second neutral point, which is located in the decaying high-energy tail of the $\sigma(E_{PE})$ curve. Table 3 shows a number of representative E_2 values collected from the literature for such materials as occur in semiconductor samples or in the electron optical column (liner tube, specimen stage, etc.). Some of the E_2 values originate from experimental measurements and others from Monte Carlo simulations.

For $E_{PE} > E_2$, the specimen becomes negatively charged because fewer SEs and back-scattered electrons are emitted than PEs hitting the target. For $E_1 < E_{PE} < E_2$, however, the specimen becomes positively charged. Local electrostatic charging of the specimen leads to unstable imaging conditions and image degradation (e.g. see Reimer, 1985). A loss of resolution occurs due to a defocussing of the PEs. However, the deflection of the PEs

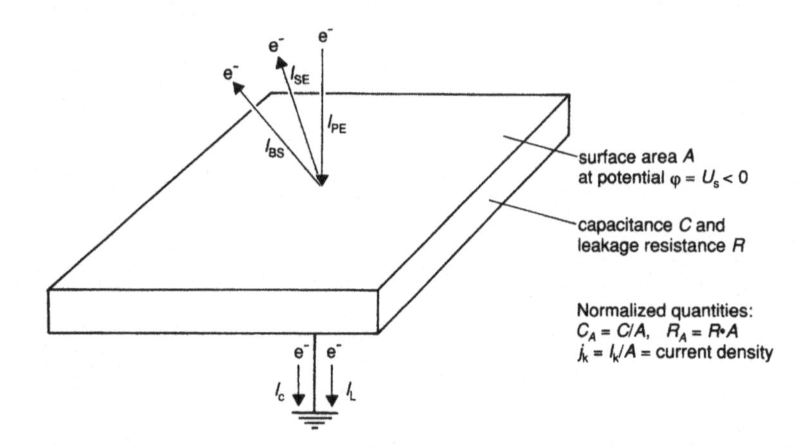

Figure 23 Currents and current densities in the case of normal and uniform irradiation of a plane specimen with a monoenergetic e-beam. I_{PE}, I_{SE}, and I_{BS} are the PE, SE, and back-scattered electron currents, respectively. I_L and I_C represent the leakage current and the capacitive charging current.

at charged specimen points can also lead to strong image distortions or to falsified line width measurements in e-beam metrology. In scanning operation, different types of charging phenomena can occur in slow scan (for image recording) and television scan (fast, for visual observation). With a real object, a discharge can occur with respect to ground (leakage current) during a scan period, so that operation is not in a steady state. The use of a thin conductive coating to avoid charging is very common in conventional scanning electron microscopy, but this method is unacceptable for in-line inspection and in-line metrology.

Let us first consider the case $E_{PE} \geqslant E_2$ and take a simple one-dimensional charging model as a basis. This analytical model from Purvis, Stevens, and Oglebay (1977) takes into account only the spatial coordinate normal to the sample surface (z-coordinate), and a monoenergetic PE beam, and describes the charging of specimens in terms of the charging of a capacitor. It is valid only for negative charging, i.e. surface potential $\varphi = U_s < 0$ (implicit sign), which occurs in the case $E_{PE} = eU_{PB} > E_2$.

From the current balance shown in Fig. 23 and using the terms introduced there, we see that the following applies for the current densities:

$$j_{PE} = j_{SE} + j_{BS} + j_L + j_C \tag{111}$$

As $j_C = -C_A dU_s/dt$ for the capacitive charging current density, we obtain the following as the charging differential equation for $U_s \doteq U_s(t)$:

$$\frac{dU_s}{dt} = \frac{-1}{C_A}\{j_{PE}(U_s) - j_{SE}(U_s) - j_{BS}(U_s) - j_L(U_s)\} \tag{112}$$

For the PE current density, Purvis et al. (1977) give

$$j_{PE}(U_s) = j_{PE}(0)\sqrt{1 + U_s/U_{PE}}$$

This relation was obtained under the assumption of constant particle density and using the law of conservation of energy. Glaser (1952), however, showed with the help of the continuity equation for the one-dimensional movement that the product of particle density and velocity and thus the current density, j_{PE}, is constant. So, with this in mind, let us modify the model of Purvis et al. (1977). For the three other current densities genuinely dependent on U_s, we then obtain

$$j_{SE}(U_s) = j_{PE}\delta(eU_{PE} + eU_s) \tag{113a}$$

$$j_{BS}(U_s) = j_{PE}\eta(eU_{PE} + eU_s) \tag{113b}$$

$$j_L(U_s) = -U_s/R_A \tag{113c}$$

and

$$R_A = RA = \sum_{i=1}^{n} \rho_i d_i \tag{113d}$$

$$C_A = \frac{C}{A} = \frac{\varepsilon_0}{\sum_{i=1}^{n} d_i/\varepsilon_i} \tag{113e}$$

In the last two equations, it was further assumed that the specimen can be composed of up to n layers, where ρ_i, d_i, and ε_i represent the specific resistance, the thickness and the relative dielectric constant of the ith layer. The uppermost layer should be so thick that it alone determines the SE yield δ and the back-scattering coefficient η. As can be seen from Eqs. (113a) and (113b), δ and η are dependent on the actual landing energy $eU_{PE} + eU_s < eU_{PE}$ at the target.

The SE yield can be described, as was done by Purvis et al. (1977), using the formula from Sternglass (1950),

$$\frac{\delta}{\delta_m} = 7.389 \frac{E_{PE}}{E_m} \exp\left(-2\sqrt{\frac{E_{PE}}{E_m}}\right) \tag{114}$$

or by using the universal yield curve (Seiler, 1983; Joy, 1987)

$$\frac{\delta}{\delta_{\mathrm{m}}} = 1.11\left(\frac{E_{\mathrm{PE}}}{E_{\mathrm{m}}}\right)^{-0.35}\left\{1 - \exp\left[-2.3\left(\frac{E_{\mathrm{PE}}}{E_{\mathrm{m}}}\right)^{1.35}\right]\right\} \tag{115}$$

Here, δ_{m} is the maximum value of δ which occurs in the case of the PE energy E_{m}. Values for these two material-dependent constants were collected from the literature for some materials and are listed in Table 3.

According to Reimer (1985), the assumption of Purvis et al. (1977) that the back-scattering coefficient is independent of the PE energy represents a good approximation in the energy range 10–100 keV. Below 5 keV, however, a stronger energy dependence occurs which can be described to a good approximation with the empirical formula

$$\eta = \eta_1 (E_{\mathrm{PE}}/\mathrm{keV})^m \tag{116}$$

with $\eta_1 \approx 0.2$, $m \approx -0.14$ for aluminum and $\eta_1 \approx 0.24$ and $m \approx -0.10$ for silicon, for example. Both η_1 and m increase with an increasing atomic number, with m having a zero crossover approximately at atomic number 30.

As the right-hand side of the differential equation for charging has a highly complex dependence on U_{s}, only a numerical solution can be considered. But if only the value assumed by U_{s} in the case of a stationary equilibrium is of interest, we obtain the following with $\mathrm{d}U_{\mathrm{s}}/\mathrm{d}t = 0$ from Eqs. (112) and (113):

$$1 + \frac{eU_{\mathrm{s}}}{eRI_{\mathrm{PE}}} = \delta(E_{\mathrm{PE}} + eU_{\mathrm{s}}) + \eta(E_{\mathrm{PE}} + eU_{\mathrm{s}}) = \sigma(E_{\mathrm{PE}} + eU_{\mathrm{s}}) \tag{117}$$

The solution of this equation can be obtained iteratively or graphically, as shown in Fig. 22, as also stated by Reimer (1985). For $R = \infty$, i.e. a floating target, we obtain directly from Eq. (117) that $E_{\mathrm{PE}} + eU_{\mathrm{s}} = E_2$ or $U_{\mathrm{s}} = -(E_{\mathrm{PE}} - E_2)/e \leqslant 0$ for $E_{\mathrm{PE}} \geqslant E_2$.

In the simple analytical charging model described above, the following aspects and cases have been neglected:

(i) The case $U_{\mathrm{s}} > 0$, which occurs when $E_1 < E_{\mathrm{PE}} < E_2$, is not covered. In this case, at least some of the emitted SEs return to the positively charged specimen. Here, the energy and angular distribution of the SEs play a role. The angular distribution satisfies Lambert's cosine law and, for the energy distribution of SEs from metals, according to

Chung and Everhart (1974), we have

$$\frac{dN_{SE}}{dE_{SE}} = k\frac{E_{SE} - E_F - \Phi_W}{(E_{SE} - E_F)^4} \tag{118}$$

Here, E_F represents the Fermi energy, Φ_W the work function and k another material constant. According to Ura et al. (1980), the following equation can also be used in the case of metals:

$$\frac{dN_{SE}}{dE_{SE}/eV} = \frac{2}{3}\frac{E_{SE}}{eV}\exp\left(2 - \sqrt{\frac{8}{3}\frac{E_{SE}}{eV}}\right) \tag{119}$$

(ii) The complete three-dimensionality of the specimen (topography, different surface materials, microfields of interconnections, etc.) and of its environment, particularly of the SE detector or of a control grid over the specimen, is not taken into account. In order to do this, a self-consistent solution of the three-dimensional Poisson equation and of the seven-dimensional Vlasov equation (collisionless Boltzmann equation) would have to be obtained for the time-dependent phase space density.

(iii) Inclination of the specimen, i.e. non-normal incidence of the PEs, is not taken into account. If the tilt angle is represented by ϑ, then, according to Seiler (1983), the following applies for $\vartheta < 80°$:

$$\delta(\vartheta) = \delta(0)/\cos^n\vartheta \tag{120}$$

The exponent $n = 1$ holds for materials with an atomic number of about 30. For light elements, n increases to about $n \approx 1.3$ and, for heavy elements, n decreases to about $n \approx 0.8$. According to Joy (1989), the following applies for the second neutral point:

$$E_2(\vartheta) = E_2(0)/\cos^2\vartheta \tag{121}$$

For this reason, specimen tilting is often used in scanning electron microscopy in order to permit charge-free operation even with energies $E_{PE} > E_2(0)$. A partial increase of E_2 also occurs in the case of rough surfaces.

(iv) Neither the conductivity nor the contamination induced by e-beams has been taken into account.

Even if contamination effects are ignored, realization of the other additional aspects mentioned in (i) to (iv) in a simulation model would present a major challenge. Rouse and Munro (1990) have simulated the SE trajectories and the SE signal for a line scan normal to a trough in negatively charged dielectric material. A negative charge fixed from the outset was assumed, and no simulation was performed of the charging process and the charge resulting from it. Nevertheless, the simulation of Rouse and Munro (1990) makes a significant contribution to understanding this type of line width measurement.

The case $E_1 < E_{PE} < E_2$, which leads to positive charging $U_s > 0$ because $\sigma > 1$, will now be briefly and rather more qualitatively considered. If the environment of the specimen is grounded, all SEs with energies $E_{SE} < eU_s$ return to the specimen. If the specimen itself is floating or insulating, i.e. $R = \infty$, then a stationary equilibrium is achieved when

$$\int_{eU_s}^{50\ \mathrm{eV}} \frac{dN_{SE}}{dE_{SE}} dE_{SE} + \eta = 1 \tag{122}$$

The positive charge is thus only a few volts and results in reduced image intensity (voltage contrast) in an SEM image. However, if a positive control grid or a collector with positive potential U_c is located directly above the specimen, the insulating surface charges up to $U_c + U_s$, i.e. a few volts more positive than the control grid.

The basic precondition for capacitive coupling voltage contrast in e-beam testing is an electron yield σ of the passivation surface greater than 1, i.e. $E_1 < E_{PE} < E_2$. Using a simple model, Görlich, Herrmann, Reiners, and Kubalek (1986) simulated the dynamic charging of 0.36 μm SiO_2 passivation over a buried interconnection line in order to explain the capacitive coupling voltage contrast. This resulted, in dependence on σ, in time constants for the attainment of an equilibrium condition of up to some tens of seconds. Fujioka, Nakamae, and Ura (1986) proposed a "core" model to explain the surface potential of passivated devices. In this model, the saddle point potential (potential barrier) above a buried electrode (interconnection line) is taken into consideration. Nye and Dinnis (1985) have investigated the influence of the extraction field on the oxide potential and the accuracy of voltage contrast measurements. Their measurements on aluminum pads (located in a previously scanned area) indicate that a high extraction field is desirable. Fujioka, Miyaji, and Ura (1988) have introduced a new method for measuring charging characteristics of electrically floating targets under e-beam irradiation by directly connecting the floating target to the gate of a MOSFET. The correct understanding of charging characteristics is of great importance for short/open tests of three-dimensional conductor networks with an e-beam (e.g. see Pfeiffer, Langner, Stickel, and Simpson, 1981 or Brunner and Lischke, 1985).

Low-voltage SEMs are increasingly used to avoid negative charging in e-beam inspection and metrology, but positive charging effects may still cause disturbance. Brunner and Menzel (1983) have investigated the influence of collector potential on surface charging. They determined the surface potential by evaluating the beam deflection of a "measuring" e-beam which passes in parallel above the target surface. They found, as already mentioned, that the surface potential assumes higher positive values when the collector potential is made more positive. According to Brunner and Menzel (1983), this can also be interpreted as an increase in the second crossover value E_2. Brunner and Schmid (1986) investigated positive charging effects in the case of e-beam metrology. They found less positive charging on insulators than on conductors. No significant measurement errors due to positive charging have been observed on insulating resist structures on semiconductor substrates with low-voltage operation. But specimens with metal structures on insulating substrates have shown measurement errors and image distortions due to positive charging.

Postek, Keery, and Larrabee (1989) have successfully applied specimen biasing to enhance or suppress SE emission from charging specimens at low acceleration voltages. Irradiation with light for the generation of photoelectrons is one method of avoiding negative charging, even in the case of higher PE electron energies. Grobman (1984) has proposed this method for testing lithographic masks with an e-beam of $E_{PE} > 5$keV. In the environmental SEM, in which the specimen is tested at about 2000 Pa, no surface charges occur because they are compensated for by the ions generated in the rough vacuum. For details about this new type of SEM, reference is made to the works of Danilatos (1988, 1990).

5. SUMMARY AND FUTURE OUTLOOK

In Part I, electron optical simulation tools have been described or cited by means of which the performance data of electron optical devices can be computed. For this purpose, the different methods of analytical and numerical computation of electrostatic or magnetic fields in guns, lenses, deflectors, analyzers, of fields in front of detectors or of microfields above the interconnection lines of ICs were first presented. The special strengths and characteristic weaknesses of different numerical field computation methods were discussed and examples of their application presented. Special attention has been drawn to the recent progress made in the development of the BEM and to the advantages of combined or hybrid methods. It was also pointed out that analytical field calculation methods may still be used to advantage for some applications.

The two main methods used to determine the electron trajectories, perturbation theory and ray tracing, have been reviewed. In the case of perturbation theory, the calculation of the fundamental and principal rays, the cardinal elements of focusing and deflection and the aberrations have been discussed. The trend is moving away from perturbation theory towards ray tracing. Various standard numerical routines can be successfully used for numerical ray tracing, but some highly accurate methods have also been developed, especially for electron optics, which deserve special attention as they are superior to general methods from the point of view of electron optics. Instead of regarding perturbation theory and ray tracing as competing methods, we should see that they complement one another. Perturbation theory has some advantages when simulating lenses and deflectors, whereas ray tracing is the method of choice for guns or SE energy analyzers and detectors. Aberrations cannot be determined only by standard perturbation theory but also by ray tracing. In the case of ray tracing, however, the most accurate method uses general rules from the perturbation Eikonal. Some fundamental laws for checking the accuracy of ray tracing, such as the conservation of the axial component of the canonical angular momentum in fields with rotational symmetry, have also been presented.

The different mathematical methods used to optimize a first electron optical design under certain constraints have been discussed relatively comprehensively. This is because the compound electrostatic and magnetic low-voltage optics for e-beam testing and electron optical inspection have more degrees of freedom than a classical magnetic objective lens of a medium-voltage SEM, so that mathematical optimization is more necessary. In the case of through-lens detection of the SEs, the objective lens must be optimized for both PE focusing and SE transport.

Coulomb interaction is the dominant factor limiting the PE beam performance in low-voltage e-beam testing and electron optical inspection. Therefore, some important formulae for the Boersch effect and the trajectory displacement effect have been collected from the literature. References have been given on Monte Carlo simulations of the Coulomb interaction, and means of reducing the Boersch effect and the trajectory displacement effect have been mentioned. Despite the great progress in the simulation of the Coulomb interaction in the last decade, further development is still necessary, especially for gun design. Since electrostatic charging of the sample is very disturbing, simulation aspects of electrostatic charging have also been treated.

Part II is entitled "Electron Optical Components and Systems". Whereas Part I has covered fundamental aspects and those relating to the calculation of fields and electron trajectories, in Part II we shall deal more directly with the hardware, i.e. the individual components such as guns, beam-blanking systems, condenser and objective lenses, deflection systems, spectrometers and detectors. We shall also consider new types of compound elements, e.g. compound spectrometer objective lens systems, and the entire electron optical column of dedicated SEM-like instruments for e-beam testing and electron optical inspection. This will all be chiefly concerned with from the aspect of low accelerating voltage.

ACKNOWLEDGMENTS

I am most obliged to my colleagues in the international electron optics community and to various publishers for permission to reproduce their published figures. Special thanks are due to Dr. B. Lencová, who made her unpublished results available to me. The sources of the figures are given in the captions. I would like to thank my Siemens colleagues Dr. T. Mertelmeier and Dr. R. Weyl and my wife Christiane for critical reading of the manuscript. I owe R. Michell my thanks for translating the manuscript. Last but not least I wish to express my gratitude to Professor T. Mulvey for encouraging this review article, for many helpful hints and suggestions and for fruitful discussion.

REFERENCES

Adams, A., & Read, F. H. (1972). *J. Phys. E, Sci. Instrum.*, *5*, 150, 156.

Adriaanse, J. P., van der Steen, H. W. G., & Barth, J. E. (1989). *J. Vac. Sci. Technol. B*, *7*, 651.

Alefeld, G., & Herzberger, J. (1974). *Einführung in die Intervallrechnung*. Mannheim: Bibliographisches Institut.

Andretta, M., Currado, A., Marini, M., & Zanarini, G. (1985). *IEEE Trans.*, *ED-32*, 983.

Argyo, W., Brust, H.-D., Fox, F., Otto, J., Plies, E., & Wolfgang, E. (1985). *Siemens Forsch. Entwickl.-Ber.*, *14*, 216.

Artsimowitsch, L. A., & Sagdejew, R. S. (1983). *Plasmaphysik für Physiker*. Stuttgart: Teubner.

Baba, N., & Kanaya, K. (1979). *J. Phys. E, Sci. Instrum.*, *12*, 525.

Beamson, G., Porter, H. Q., & Turner, D. W. (1980). *J. Phys. E, Sci. Instrum.*, *13*, 64.

Becker, R. (1989). *Nucl. Instr. Methods B*, *42*, 303.

Becker, R. (1990). *Nucl. Instr. Methods A*, *298*, 13.

Bellman, R. E. (1957). *Dynamic programming*. Princeton: Princeton University Press.

Berezin, I. S., & Zhidkov, N. P. (1965). *Computing methods, vol. II*. Oxford: Pergamon Press.
Berger, A., Spehr, R., & Rose, H. (1990). *Optik, 86*, 77.
Bernhard, W. (1980). *Optik, 57*, 73.
Bertram, S. (1940). *Proc. Inst. Radio Eng., 28*, 418.
Bertram, S. (1942). *J. Appl. Phys., 13*, 496.
Berz, M. (1987). *Nucl. Instrum. Methods A, 25*, 431.
Berz, M. (1988). *IEEE Trans., ED-35*, 2002.
Berz, M. (1990). *Nucl. Instrum. Methods A, 298*, 426.
Betz, A. (1964). *Konforme Abbildung* (2nd ed.). Berlin: Springer-Verlag.
Bialy, H., & Olbrich, M. (1975). *Optimierung*. Leipzig: VEB Fachbuchverlag.
Billat, S. P. (1988). In *Technical proceedings of SEMICON Europa, Zürich*. Semiconductor Equipment and Materials International (p. 109).
Binns, K. J., & Lawrenson, P. J. (1973). *Analysis and computation of electric and magnetic field problems* (2nd ed.). Oxford: Pergamon Press.
Birtles, A. B., Mayo, B. J., & Bennet, A. W. (1973). *Proc. IEEE, 120*, 213.
Boersch, H. (1954). *Z. Phys., 139*, 115.
Bonjour, P. (1980). In A. Septier (Ed.), *Applied charged particle optics, part A* (p. 1). New York: Academic Press.
Born, M. (1933). *Optik*. Berlin: Springer-Verlag.
Born, M., & Wolf, E. (1985). *Principles of optics* (6th ed.). Oxford: Pergamon Press.
Brüche, E., & Scherzer, O. (1934). *Geometrische Elektronenoptik*. Berlin: Springer-Verlag.
Brunner, M., & Lischke, B. (1985). *Scanning Electron Microsc., III*, 991.
Brunner, M., & Menzel, E. (1983). *J. Vac. Sci. Technol. B, 1*, 1344.
Brunner, M., & Schmid, R. (1986). *Scanning Electron Microsc., II*, 377.
Buchholz, H. (1957). *Elektrische und Magnetische Potentialfelder*. Berlin: Springer-Verlag.
Bulirsch, R., & Stoer, J. (1966). *Numer. Math., 8*, 1.
Cap, F. (1975). *Einführung in die Plasmaphysik I, Theoretische Grundlagen* (2nd ed.). Braunschweig: Vieweg.
Chu, H. C., & Munro, E. (1981). *J. Vac. Sci. Technol., 19*, 1053.
Chu, H. C., & Munro, E. (1982a). *Optik, 61*, 121.
Chu, H. C., & Munro, E. (1982b). *Optik, 61*, 213.
Chung, M. S., & Everhart, T. E. (1974). *J. Appl. Phys., 45*, 707.
Collatz, L. (1960). *The numerical treatment of differential equations* (3rd ed.). Berlin: Springer-Verlag.
Courant, R. (1943). *Bull. Am. Math. Soc., 49*, 1.
Cruise, D. R. (1963). *J. Appl. Phys., 34*, 3477.
Danilatos, G. D. (1988). *Adv. Electron. Electron Phys., 71*, 109.
Danilatos, G. D. (1990). *Scanning, 12*, 23.
Davidson, M., Kaufman, K., Mazor, I., & Cohen, F. (1987). *Proc. SPIE, 775*, 233.
Dekkers, N. H. (1974). *J. Phys. D, Appl. Phys., 7*, 805.
Dennis, J. E. (1984). *Proc. IEEE, 72*, 1765.
Dinnis, A. R. (1987). *Microelectron. Eng., 7*, 139.
Dinnis, A. R. (1988). *Scanning Microsc., 2*, 1407.
Dinnis, A. R., & Khursheed, A. (1989). *Microelectron. Eng., 9*, 445.
Dommaschk, W. (1965). *Optik, 23*, 472.
Dubbeldam, L. (1989). *A voltage contrast detector with double channel energy analyzer in a scanning electron microscope* (Ph.D. thesis). Delft University of Technology.
Dugas, J., Durandeau, P., & Fert, C. (1961). *Rev. Opt., 40*, 277.

Durand, E. (1964). *Électrostatique, tome 1*. Paris: Masson.
Durand, E. (1966a). *Électrostatique, tome 2*. Paris: Masson.
Durand, E. (1966b). *Électrostatique, tome 3*. Paris: Masson.
Durand, E. (1968). *Magnétostatique*. Paris: Masson.
Durandeau, P. (1957). *Ann. Fac. Sci. Univ. Toulouse Sci. Math. Sci. Phys.*, *21*, 1.
Durandeau, P., & Fert, C. (1957). *Rev. Opt.*, *36*, 205.
Dyke, W. P., Trolan, J. K., Dolan, W., & Barnes, G. (1953). *J. Appl. Phys.*, *24*, 570.
El-Kareh, A. B., & El-Kareh, J. C. J. (1970a). *Electron beams, lenses and optics, vol. 1*. New York: Academic Press.
El-Kareh, A. B., & El-Kareh, J. C. J. (1970b). *Electron beams, lenses and optics, vol. 2*. New York: Academic Press.
El-Kareh, A. B., & Smither, M. A. (1979). *J. Appl. Phys.*, *50*, 5596.
Enright, W. H., & Hull, T. E. (1976). *SIAM J. Numer. Anal.*, *13*, 944.
Eupper, M. (1982). *Optik*, *62*, 299.
Franzen, N. (1984). In J. J. Hren, F. A. Lenz, E. Munro, & P. B. Sewell (Eds.), *Electron optical systems for microscopy, microanalysis and microlithography* (p. 115). O'Hare: Scanning Electron Microscopy.
Franzen, N., & Munro, E. (1987). In *Proc. ISEOB 86 (Beijing)* (p. 57).
Frobin, W. (1968). *Optik*, *27*, 203.
Frosien, J., & Plies, E. (1987). *Microelectron. Eng.*, 7, 163.
Frosien, J., Plies, E., & Anger, K. (1989). *J. Vac. Sci. Technol. B*, 7, 1874.
Fujioka, H., Miyaji, K., & Ura, K. (1988). *J. Phys. E, Sci. Instrum.*, *21*, 583.
Fujioka, H., Nakamae, K., & Ura, K. (1986). In *Proc. 11th int. congr. electron microsc. (Kyoto), vol. 1* (p. 643).
Garth, S. C. J., Nixon, W. C., & Spicer, D. F. (1986). *J. Vac. Sci. Technol. B*, *4*, 217.
Gill, P. E., Murray, W., & Wright, M. H. (1981). *Practical optimization*. London: Academic Press.
Ginsberg, D. M., & Melchner, M. J. (1970). *Rev. Sci. Instrum.*, *41*, 122.
Glaser, W. (1933). *Z. Phys.*, *80*, 451.
Glaser, W. (1941). *Z. Phys.*, *117*, 285.
Glaser, W. (1949). *Ann. Phys. (Leipz.)*, *4*, 389.
Glaser, W. (1952). *Grundlagen der Elektronenoptik*. Wien: Springer-Verlag.
Glaser, W. (1956). In S. Flügge (Ed.), *Handbuch der Physik, vol. 33* (p. 123). Berlin: Springer-Verlag.
Glaser, W., & Lenz, F. (1951). *Ann. Phys. (Leipz.)*, *9*, 19.
Glaser, W., & Schiske, P. (1953). *Z. Angew. Phys.*, *5*, 329.
Glaser, W., & Schiske, P. (1954a). *Optik*, *11*, 422.
Glaser, W., & Schiske, P. (1954b). *Optik*, *11*, 445.
Glaser, W., & Schiske, P. (1955). *Optik*, *12*, 233.
Glatzel, E. (1961). *Optik*, *18*, 577.
Glatzel, U., & Lenz, F. (1988). *Optik*, *79*, 15.
Glikman, L. G., Kel'man, V. M., & Nurmanov, M. Sh. (1974). *Sov. Phys. Tech. Phys.*, *18*, 864.
Gopinath, A. (1987). *Adv. Electron. Electron Phys.*, *69*, 1.
Gopinath, A. (1989). In N. G. Einspruch, S. S. Cohen, & R. N. Singh (Eds.), *VLSI electronics: microstructure science, vol. 21* (p. 477). San Diego: Academic Press.
Görlich, S., Herrmann, K. D., Reiners, W., & Kubalek, E. (1986). *Scanning Electron Microsc.*, *II*, 447.

Goto, E., & Soma, T. (1977). *Optik, 48*, 255.
Gray, F. (1939). *Bell Syst. Tech. J., 18*, 1.
Grivet, P. (1951). *C. R. Acad. Sci. Paris, 233*, 921.
Grivet, P. (1952). *C. R. Acad. Sci. Paris, 234*, 73.
Grivet, P. (1972). *Electron optics* (2nd revised English ed.). Oxford: Pergamon Press.
Grobman, W.D. (1984). Electron beam system with reduced charge buildup. US Patent 4,453,086, priority date Dec. 31, 1981.
Groves, T., Hammond, D. L., & Kuo, H. (1979). *J. Vac. Sci. Technol., 16*, 1680.
Grümm, H., & Spurny, H. (1956). *Österr. Ing. Arch., 10*, 104.
Gu, C.-x., & Shan, L.-y. (1984). In J. J. Hren, F. A. Lenz, E. Munro, & P. B. Sewell (Eds.), *Electron optical systems for microscopy, microanalysis and microlithography* (p. 91). O'Hare: Scanning Electron Microscopy.
Haantjes, J., & Lubben, G. J. (1957). *Philips Res. Rep., 12*, 46.
Haantjes, J., & Lubben, G. J. (1959). *Philips Res. Rep., 14*, 65.
Hahn, E. (1958a). *Jena. Jahrb., I*, 184.
Hahn, E. (1958b). *Optik, 15*, 500.
Hahn, E. (1959a). *Jena. Jahrb., I*, 86.
Hahn, E. (1959b). *Optik, 16*, 513.
Hahn, E. (1985). *Optik, 69*, 45.
Hamming, R. W. (1959). *J. Assoc. Comput. Mach., 6*, 37.
Hamming, R. W. (1973). *Numerical methods for scientists and engineers* (2nd ed.). New York: McGraw-Hill.
Hanssum, H. (1984). *J. Phys. D, Appl. Phys., 17*, 1.
Hanssum, H. (1985). *J. Phys. D, Appl. Phys., 18*, 1971.
Hanssum, H. (1986). *J. Phys. D, Appl. Phys., 19*, 493.
Harrington, R. F. (1968). *Field computation by moment methods* (Chap. 2), New York: Macmillan.
Harting, E., & Read, F. H. (1976). *Electrostatic lenses*. Amsterdam: Elsevier.
Hawkes, P. W. (1966). *Quadrupole optics*. Berlin: Springer-Verlag.
Hawkes, P. W. (1970). *Quadrupoles in electron lens design*. New York: Academic Press.
Hawkes, P. W. (1973). *Comput. Aided Des., 5*, 200.
Hawkes, P. W. (1980). In A. Septier (Ed.), *Applied charged particle optics, part A* (p. 45). New York: Academic Press.
Hawkes, P. W. (1982). In P. W. Hawkes (Ed.), *Magnetic electron lenses* (p. 413). Berlin: Springer-Verlag.
Hawkes, P. W. (1989). *Optik, 83*, 104, 122.
Hawkes, P. W., & Kasper, E. (1989a). *Principles of electron optics, vol. 1*. London: Academic Press.
Hawkes, P. W., & Kasper, E. (1989b). *Principles of electron optics, vol. 2*. London: Academic Press.
Heijnemans, W. A. L., Nieuwendijk, J. A. M., & Vink, N. G. (1980/1981). *Philips Techn. Rdsch., 39*, 158.
Herrmann, K.-H., Menadue, J., & Pearce-Percy, H. T. (1976). In *Proc. 6th Europ. congr. electron microsc. (Jerusalem)* (p. 342).
Heun, K. (1889). *Math. Ann., 33*, 161.
Hildebrandt, H.-J. (1954). *Zur Dimensionierung des Eisenkreises elektromagnetischer Elektronenlinsen* (Diploma-thesis). Free University of Berlin.
Hildebrandt, H.-J., & Riecke, W. D. (1966). *Z. Angew. Phys., 20*, 336.

Hoch, H., Kasper, E., & Kern, D. (1978). *Optik, 50*, 413.
Huchital, D. A., & Rigden, J. D. (1972). *J. Appl. Phys., 43*, 2291.
Hutter, R. G. (1945). *J. Appl. Phys., 16*, 678.
Hutter, R. (1967). In A. Septier (Ed.), *Focusing of charged particles, vol. 2* (p. 3). New York: Academic Press.
Hutter, R. G. E. (1974). In B. Kazan (Ed.), *Advances in image pickup and display, vol. 1* (p. 163). New York: Academic Press.
Idesawa, M., Soma, T., Goto, E., & Sasaki, T. (1983). *J. Vac. Sci. Technol. B, 1*, 1322.
Iselin, C. F. (1981). *IEEE Trans., MAG-17*, 2168.
Ishii, T., Hyozo, M., Tada, T., & Maki, N. (1989). In *Proc. symp. electron beam testing (Osaka)* (p. 109).
Jackson, J. D. (1975). *Classical electrodynamics* (2nd ed.). New York: Wiley.
Janse, J. (1971). *Optik, 33*, 270.
Jansen, G. H. (1988a). *Coulomb interactions in particle beams* (Ph.D. thesis). Delft University of Technology.
Jansen, G. H. (1988b). *J. Vac. Sci. Technol. B, 6*, 1977.
Jansen, G. H. (1990). *Adv. Electron. Electron Phys., Suppl., 21*.
Janzen, R. (1990). *Computersimulation von Elektronenstrahltest-Spektrometern* (Diploma thesis). Technical University of Darmstadt.
Jordan-Engeln, G., & Reutter, F. (1972). *Numerische Mathematik für Ingenieure*. Mannheim: Bibliographisches Institut.
Joy, D. C. (1987). *J. Microsc., 147*(Pt 1), 51.
Joy, D. C. (1989). *Scanning, 11*, 1.
Juma, S. M. (1986). *J. Phys. E, Sci. Instrum., 19*, 457.
Kaashoek, J. (1968). *Philips Res. Rep., Suppl., 11*, 1.
Kanaya, K., & Baba, N. (1977). *Optik, 47*, 239.
Kanaya, K., & Baba, N. (1978). *J. Phys. E, Sci. Instrum., 11*, 265.
Kanaya, K., Baba, N., & Ono, S. (1976). *Optik, 46*, 125.
Kanaya, K., Kawakatsu, H., & Miya, T. (1972). *J. Electron Microsc., 21*, 261.
Kanaya, K., Kawakatsu, H., Yamazaki, H., & Sibata, S. (1966). *J. Sci. Instrum., 43*, 416.
Kang, N. K., Orloff, J., Swanson, L. W., & Tuggle, D. (1981). *J. Vac. Sci. Technol., 19*, 1077.
Kang, N. K., Tuggle, D., & Swanson, L. W. (1983). *Optik, 63*, 313.
Kasper, E. (1976). *Optik, 46*, 271.
Kasper, E. (1979). *Optik, 54*, 135.
Kasper, E. (1981). *Nucl. Instrum. Methods, 187*, 175.
Kasper, E. (1982). In P. W. Hawkes (Ed.), *Magnetic electron lenses* (p. 57). Berlin: Springer-Verlag.
Kasper, E. (1984a). In J. J. Hren, F. A. Lenz, E. Munro, & P. B. Sewell (Eds.), *Electron optical systems for microscopy, microanalysis and microlithography* (p. 63). O'Hare: Scanning Electron Microscopy.
Kasper, E. (1984b). *Optik, 68*, 341.
Kasper, E. (1985). *Optik, 69*, 117.
Kasper, E. (1987). *Nucl. Instrum. Methods A, 258*, 466.
Kasper, E. (1990). *Nucl. Instrum. Methods A, 298*, 295.
Kasper, E., & Lenz, F. (1980). In *Proc. 7th Europ. congr. electron microsc (The Hague), vol. 1* (p. 10).
Kasper, E., & Scherle, W. (1982). *Optik, 60*, 339.
Kasper, E., & Ströer, M. (1989). *Optik, 83*, 93.

Kasper, E., & Ströer, M. (1990). *Nucl. Instrum. Methods A, 298*, 1.
Kelly, J. (1977). *Adv. Electron. Electron Phys., 43*, 43.
Kern, D. (1978). *Theoretische Untersuchungen an rotations-symmetrischen Strahlerzeugungssystemen mit Feldemissionsquelle* (Ph.D. thesis). University of Tübingen.
Kern, D. P. (1979). *J. Vac. Sci. Technol., 16*, 1686.
Khursheed, A., & Dinnis, A. R. (1989). *J. Vac. Sci. Technol. B*, 7, 1882.
Killes, P. (1985). *Optik, 70*, 64.
Klemperer, O. (1953). *Electron optics* (2nd ed.). Cambridge: Cambridge University Press.
Klemperer, O. (1972). *Electron physics, the physics of the free electron* (2nd ed.). London: Butterworth.
Knauer, W. (1981). *Optik, 59*, 335.
Knoll, M. (1935). *Z. Tech. Phys., 11*, 467.
Kramer, J. (1967). *Br. J. Appl. Phys., 18*, 1815.
Kruit, P., & Dubbeldam, L. (1987). *Scanning Microsc., 1*, 1641.
Kruit, P., & Read, F. H. (1983). *J. Phys. E, Sci. Instrum., 16*, 313.
Kunze, D., & Janssen, D. (1987). *Optik, 76*, 61.
Kunze, D., & Janssen, D. (1989). *Optik, 82*, 132.
Künzi, H. P., Krelle, W., & von Randow, R. (1979). *Nichtlineare Programmierung* (2nd ed.). Berlin: Springer-Verlag.
Künzi, H. P., Tzschach, H. G., & Zehnder, C. A. (1967). *Numerische Methoden der mathematischen Optimierung*. Stuttgart: Teubner.
Küpfmüller, K. (1965). *Einführung in die theoretische Elektrotechnik* (8th ed.). Berlin: Springer-Verlag.
Kuroda, K. (1983). *Optik, 64*, 125.
Langmuir, D. B. (1937). *Proc. Inst. Radio Eng., 25*, 977.
Lawson, J. D. (1977). *The physics of charged-particle beams*. Oxford: Clarendon Press.
Lenc, M., & Lencová, B. (1988). *Optik*, 7, 127.
Lencová, B. (1988a). *Optik, 79*, 1.
Lencová, B. (1988b). In *Proc. EUREM 88 (York, UK), vol. 1* (p. 75).
Lencová, B., & Lenc, M. (1982). In *Proc. 10th int. congr. electron microsc. (Hamburg), vol. 1* (p. 317).
Lencová, B., & Lenc, M. (1984). *Optik, 68*, 37.
Lencová, B., & Lenc, M. (1986). *Scanning Electron Microsc., III*, 897.
Lencová, B., & Lenc, M. (1989). *Optik, 82*, 64.
Lencová, B., Lenc, M., & van der Mast, K. D. (1989). *J. Vac. Sci. Technol. B*, 7, 1846.
Lenz, F. (1950a). *Optik*, 7, 243.
Lenz, F. (1950b). *Z. Angew. Phys., 2*, 448.
Lenz, F. (1951). *Ann. Phys. (Leipz.), 9*, 245.
Lenz, F. (1956a). *Z. Angew. Phys., 8*, 492.
Lenz, F. (1956b). *Ann. Phys. (Leipz.), 19*, 82.
Lenz, F. (1982). In P. W. Hawkes (Ed.), *Magnetic electron lenses* (p. 117). Berlin: Springer-Verlag.
Levenberg, K. (1944). *Q. Appl. Math., 2*, 164.
Li, Y. (1983). *Optik, 63*, 213.
Liebmann, G. (1951). *Proc. Phys. Soc. B, 64*, 972.
Liebmann, G., & Grad, L. M. (1951). *Proc. Phys. Soc. B, 64*, 956.
Liebmann, H. (1918). *Sitzungsber. Math.-Naturwiss. Kl. Bayer. Akad. Wiss. München*, 385.
Lischke, B., Plies, E., & Schmitt, R. (1983). *Scanning Electron Microsc., III*, 1177.

Loeffler, K. H., & Hudgin, R. H. (1970). In *Proc. 7th int. congr. electron microsc. (Grenoble), vol. 2* (p. 67).
Lucas, I. (1976). *J. Appl. Phys.*, *47*, 1645.
Menzel, E., & Kubalek, E. (1983). *Scanning*, *5*, 151.
Mills, F. E., & Morgan, G. H. (1972). *Part. Accel.*, *5*, 227.
Moon, P., & Spencer, D. E. (1961). *Field theory handbook*. Berlin: Springer-Verlag.
Moore, R. E. (1966). *Interval analysis*. Englewood-Cliffs: Prentice-Hall.
Morrison, P. (1961). In S. Flügge (Ed.), *Encyclopedia of physics, vol. 46/1* (p. 1). Berlin: Springer-Verlag.
Moses, R. W. (1973). In P. W. Hawkes (Ed.), *Image processing and computer-aided design in electron optics* (p. 250). London: Academic Press.
Mulvey, T. (1975). Magnetic lenses. US Patent 3,870,891, priority date Sept. 4, 1972.
Mulvey, T. (1982). In P. W. Hawkes (Ed.), *Magnetic electron lenses* (p. 359). Berlin: Springer-Verlag.
Mulvey, T. (1984). In J. J. Hren, F. A. Lenz, E. Munro, & P. B. Sewell (Eds.), *Electron optical systems for microscopy, microanalysis and microlithography* (p. 15). O'Hare: Scanning Electron Microscopy.
Mulvey, T., & Nasr, H. (1981). *Nucl. Instrum. Methods*, *187*, 201.
Mulvey, T., & Wallington, M. J. (1973). *Rep. Prog. Phys.*, *36*, 347.
Munack, H. (1990). *Optik*, *85*, 161.
Munack, H. (1991). *Globale Optimierung elektronenoptischer Systeme mittels Intervallarithmetik* (Ph.D. thesis). University of Tübingen.
Munro, E. (1970). In *Proc. 7th int. congr. electron microsc. (Grenoble), vol. 2* (p. 55).
Munro, E. (1971). *Computer-aided design methods in electron optics* (Ph.D. thesis). Cambridge University.
Munro, E. (1973). In P. W. Hawkes (Ed.), *Image processing and computer-aided design in electron optics* (p. 284). London: Academic Press.
Munro, E. (1974). *Optik*, *39*, 450.
Munro, E. (1975). *J. Vac. Sci. Technol.*, *12*, 1146.
Munro, E. (1980). In A. Septier (Ed.), *Applied charged particle optics, part B* (p. 73). New York: Academic Press.
Munro, E. (1987a). In *Proc. ISEOB 86 (Beijing)* (p. 177).
Munro, E. (1987b). *Nucl. Instrum. Methods A*, *258*, 443.
Munro, E. (1988). *J. Vac. Sci. Technol. B*, *6*, 941.
Munro, E., & Chu, H. C. (1982a). *Optik*, *60*, 371.
Munro, E., & Chu, H. C. (1982b). *Optik*, *61*, 1.
Munro, E., & Wittels, N. D. (1977). *Optik*, *47*, 25.
Muray, A., & Richardson, N. (1987). In *Proc. symp. electron beam testing (Osaka)* (p. 101).
Nagatani, T., Saito, S., Sato, M., & Yamada, M. (1987). *Scanning Microsc.*, *1*, 901.
Nagy, Gy. A., & Szilagyi, M. (1974). *Introduction to the theory of space-charge optics* (2nd ed.). New York/London: Halsted Press/Macmillan.
Nakamae, K., Fujioka, H., & Ura, K. (1981). *J. Phys. D, Appl. Phys.*, *14*, 1939.
Nerstheimer, K. (1979). *VDI Nachr.*, *5*, 8.
Numerov, B. (1933). *Publ. J. Observ. Astrophys. Central Russie*, *2*, 188.
Nye, P., & Dinnis, A. (1985). *Scanning*, *7*, 117.
Oatley, C. W., & Everhart, T. E. (1957). *J. Electron.*, *2*, 568.
Ohiwa, H. (1977). *J. Phys. D, Appl. Phys.*, *10*, 1437.
Ohiwa, H. (1978). *J. Vac. Sci. Technol.*, *15*, 849.

Ohiwa, H. (1985). *Optik, 70*, 72.
Ohiwa, H., Goto, E., & Ono, A. (1971). *Electron. Commun. Jpn., 54-B*, 44.
Ohtaka, T., Saito, S., Furuya, T., & Yamada, O. (1985). *Proc. SPIE, 565*, 205.
Okayama, S. (1990). *Nucl. Instrum. Methods A, 298*, 488.
Oku, K., & Fukushima, M. (1986). *IEEE Trans., ED-33*, 1090.
Ollendorf, F. (1932). *Potentialfelder der Elektrotechnik*. Berlin: Springer-Verlag.
Ollendorf, F. (1952). *Berechnung magnetischer Felder. Tech. Elektrodyn., Band 1*. Wien: Springer.
Pawley, J. B. (1990). *Scanning, 12*. I-22.
Pease, R. F. W. (1967). In *Proceedings of the 9th symposium on electron, ion and laser beam technology* (p. 176). San Francisco: San Francisco Press.
Pfeiffer, H. C., Langner, G. O., Stickel, W., & Simpson, R. A. (1981). *J. Vac. Sci. Technol., 19*, 1014.
Picht, J. (1957). *Einführung in die Theorie der Elektronenoptik* (2nd ed.). Leipzig: J.A. Barth.
Pierce, J. R. (1949). *Theory and design of electron beams*. Princeton: Van Nostrand.
Plies, E. (1973). *Optik, 38*, 502.
Plies, E. (1982a). *Siemens Forsch. Entwickl.-Ber., 11*, 38.
Plies, E. (1982b). *Siemens Forsch. Entwickl.-Ber., 11*, 83.
Plies, E. (1990a). *Microelectron. Eng., 12*, 89.
Plies, E. (1990b). *Nucl. Instrum. Methods A, 298*, 142.
Plies, E., & Elstner, R. (1989a). *Optik, 82*, 57.
Plies, E., & Elstner, R. (1989b). *Optik, 83*, 11.
Plies, E., & Otto, J. (1985). *Scanning Electron Microsc., IV*, 1491.
Plies, E., & Schweizer, M. (1987). *Siemens Forsch. Entwickl.-Ber., 16*, 30.
Plies, E., & Typke, D. (1978). *Z. Naturforsch., 33a*, 1361. Erratum: *Z. Naturforsch., 35a*, 566.
Postek, M. T., Keery, W. J., & Larrabee, R. D. (1989). *Scanning, 11*, 111.
Powell, M. J. D. (1965). *Comput. J.*, 7, 303.
Preikszas, D. (1990). *Optimierung rotationssymmetrischer, kombiniert elektro-magnetischer Elektronenlinsen* (Diploma-thesis). Technical University of Darmstadt.
Purvis, C. K., Stevens, N. J., & Oglebay, J. C. (1977). *Charging characteristics of materials: Comparison of experimental results with simple analytical models*. NASA Technical Memorandum X-73606.
Ralston, A. (1972). In A. Ralston, & H. S. Wilf (Eds.), *Mathematische Methoden für Digitalrechner I* (2nd ed.) (p. 169). München: Oldenbourg.
Rao, S. (1984). *Optimization, theory and applications*. New Delhi: Halstead–Wiley.
Rauh, H. (1971). *Z. Naturforsch., 26a*, 1667.
Read, F. H. (1971). *J. Phys. E, Sci. Instrum., 4*, 562.
Read, F. H., Adams, A., & Soto-Montiel, J. R. (1971). *J. Phys. E, Sci. Instrum., 4*, 625.
Rechenberg, I. (1973). *Evolutionsstrategie-Optimierung technischer Systeme nach Prinzipien der biologischen Evolution*. Stuttgart: Fromann–Holzboog.
Recknagel, A. (1941). *Z. Phys., 117*, 689.
Regenstreif, E. (1951). *Ann. Radioélectr., 6*(51), 114.
Reimer, L. (1985). *Scanning electron microscopy*. Berlin: Springer-Verlag.
Richardson, J. A., & Kuester, J. L. (1973). *Commun. ACM, 16*, 487.
Riecke, W. D. (1982). In P. W. Hawkes (Ed.), *Magnetic electron lenses* (p. 163). Berlin: Springer-Verlag.
Ritz, E. F. (1979). *Adv. Electron. Electron Phys., 59*, 299.
Romberg, W. (1955). *Det. Kong. Norske Vidensk. Selsk. Forh., 28*(7).

Rose, H. (1966/1967). *Optik, 24*, 36.
Rose, H. (1968). *Optik, 27*, 466.
Rose, H. (1978). *Optik, 51*, 15.
Rose, H. (1987). *Nucl. Instrum. Methods A, 258*, 374.
Rose, H., & Moses, R. W. (1973). *Optik, 37*, 316.
Rose, H., & Spehr, R. (1980). *Optik, 57*, 339.
Rose, H., & Spehr, R. (1983). In A. Septier (Ed.), *Applied charged particle optics, part C* (p. 475). New York: Academic Press.
Rouse, J., & Munro, E. (1989). *J. Vac. Sci. Technol. B*, 7, 1891.
Rouse, J., & Munro, E. (1990). *Nucl. Instrum. Methods A, 298*, 78.
Ruska, E. (1965a). *J. R. Microsc. Soc., 87*, 77.
Ruska, E. (1965b). *Optik, 22*, 319.
Sasaki, T. (1979). In *Proc. VLSI conf. 1979* (p. 125).
Sasaki, T. (1982). *J. Vac. Sci. Technol., 21*, 695.
Sasaki, T. (1986). *J. Vac. Sci. Technol. B, 4*, 135.
Scheinfein, M., & Galantai, A. (1986). *Optik, 74*, 154.
Scherle, W. (1983). *Optik, 63*, 217.
Scherzer, O. (1936). *Z. Phys., 101*, 593.
Scherzer, O. (1937). In H. Busch, & E. Brüche (Eds.), *Beiträge zur Elektronenoptik* (p. 33). Leipzig: J.A. Barth.
Scherzer, O. (1971). *Optik, 33*, 501.
Scherzer, O. (1980). *Optik, 56*, 133.
Schlesinger, K. (1952). *Electronics, 25*, 105.
Schlesinger, K. (1956). *Proc. Inst. Radio Eng., 44*, 659.
Schönecker, G., Rose, H., & Spehr, R. (1986). *Ultramicroscopy, 20*, 203.
Schönecker, G., Spehr, R., & Rose, H. (1990). *Nucl. Instrum. Methods A, 298*, 360.
Schwefel, H.-P. (1977). *Numerische Optimierung von Computermodellen mittels Evolutionsstrategien*. Basel: Birkhäuser.
Schwertfeger, W., & Kasper, E. (1974). *Optik, 41*, 160.
Seiler, H. (1983). *J. Appl. Phys., 54*, R1.
Shao, Z. (1989a). *Rev. Sci. Instrum., 60*, 693.
Shao, Z. (1989b). *Rev. Sci. Instrum., 60*, 3434.
Shao, Z., & Crewe, A. V. (1988). *Optik, 79*, 105.
Sheppard, C. J. R., & Ahmed, H. (1976). *Optik, 44*, 139.
Shimada, H., Mimura, R., Sawaragi, H., Suzuki, Y., & Aihara, R. (1986). In *Proc. 11th int. congr. electron microsc. (Kyoto), vol. 1* (p. 385).
Silvester, P., & Konrad, A. (1973). *Int. J. Numer. Methods Eng., 5*, 481.
Simkin, J., & Trowbridge, C. V. (1976). In *Proc. COMPUMUAG (Oxford)* (p. 5).
Simonyi, K. (1966). *Theoretische Elektrotechnik* (2nd revised ed.). Berlin: Deutscher Verlag der Wissenschaften.
Simpson, J. A. (1961). *Rev. Sci. Instrum., 32*, 1283.
Singer, B., & Braun, M. (1970). *IEEE Trans., ED-17*, 926.
Smith, M. R., & Munro, E. (1986). *Optik, 74*, 7.
Smythe, W. R. (1950). *Static and dynamic electricity* (2nd ed.). New York: McGraw Hill.
Soma, T. (1977). *Optik, 49*, 255.
Sommerfeld, A. (1964). *Mechanik* (7th ed.). Leipzig: Akademie Verlagsgesellschaft, Geest und Portig.
Sternglass, E. J. (1950). *Phys. Rev., 80*, 925.

Stoer, J., & Bulirsch, R. (1990). *Numerische Mathematik 2* (3rd ed.). Berlin: Springer-Verlag.
Ströer, M. (1987). *Optik*, *77*, 15.
Ströer, M. (1988). *Optik*, *81*, 13.
Sturans, M. A., & Pfeiffer, H. C. (1983). In *Proc. microcircuit eng. 83 (Cambridge)* (p. 107).
Sturrock, P. (1951). *Philos. Trans. R. Soc. Lond. A*, *243*, 387.
Szilagyi, M. (1977). *Optik*, *48*, 215. *Optik*, *49*, 223.
Szilagyi, M. (1983). *J. Vac. Sci. Technol. B*, *1*, 1137.
Szilagyi, M. (1988). *Electron and ion optics*. New York: Plenum Press.
Szilagyi, M., Yakowitz, S., & Duff, M. (1984). *Appl. Phys. Lett.*, *44*, 7.
Tahir, K., & Mulvey, T. (1990). *Nucl. Instrum. Methods A*, *298*, 389.
Tretner, W. (1954). *Optik*, *11*, 312. Erratum: *Optik*, *12*, 293.
Tretner, W. (1956). *Optik*, *13*, 516.
Tretner, W. (1959). *Optik*, *16*, 155.
Typke, D. (1972). *Optik*, *34*, 573.
Ura, K. (1981). *Optik*, *58*, 281.
Ura, K., & Fujioka, H. (1989). *Adv. Electron. Electron Phys.*, *73*, 233.
Ura, K., Fujioka, H., & Yokobayashi, T. (1980). In *Proc. 7th Europ. congr. electron microsc. (The Hague), vol. 1* (p. 330).
Urankar, L. K. (1980). *IEEE Trans.*, *MAG-16*, 1283.
van der Merwe, J. P. (1978). *J. Appl. Phys.*, *49*, 4335.
van der Merwe, J. P. (1979). *J. Appl. Phys.*, *50*, 2506.
van der Steen, H. W. G., & Barth, J. E. (1989). *J. Vac. Sci. Technol. B*, 7, 1886.
van Hoof, H. A. (1980). *J. Phys. E, Sci. Instrum.*, *13*, 1081.
van Leeuwen, J. M. J., & Jansen, G. H. (1983). *Optik*, *65*, 179.
van Ments, M., & Le Poole, J. B. (1947). *Appl. Sci. Res.*, *1B*, 3.
von Ardenne, M. (1938a). *Z. Phys.*, *109*, 553.
von Ardenne, M. (1938b). *Z. Tech. Phys.*, *19*, 407.
von Koppenfels, W., & Stallmann, F. (1959). *Praxis der konformen Abbildung*. Berlin: Springer-Verlag.
Walsh, J. (1977). In D. Jacobs (Ed.), *The state of art in numerical analysis* (p. 501). London: Academic Press.
Weber, C. (1967). In A. Septier (Ed.), *Focusing of charged particles, vol. 1* (p. 45). New York: Academic Press.
Weber, E. (1950). *Electromagnetic fields, vol. 1*. New York: Wiley.
Weidenhausen, A., & Spehr, R. (1989). *Optik*, *83*, 55.
Weidenhausen, A., Spehr, R., & Rose, H. (1985). *Optik*, *69*, 126.
Weidlich, E.-R. (1990). *Microelectron. Eng.*, *11*, 347.
Wendt, G. (1939). *Telefunkenröhre*, *15*, 100.
Wendt, G. (1948). *Ann. Phys.*, *2*, 256.
Wendt, G. (1954). *Ann. Radioélectr.*, *9*, 286.
Wendt, G. (1958). In S. Flügge (Ed.), *Handbuch der Physik, Band 16* (p. 1). Berlin: Springer-Verlag.
Wijnaendts van Resandt, R. W., & Zapf, Th. (1988). In *Technical proceedings of SEMICON Europa, Zürich* (p. 129). Semiconductor Equipment and Materials International.
Wolfgang, E. (1986). *Microelectron. Eng.*, *4*, 77.
Wollnik, H. (1987). *Optics of charged particles*. Orlando: Academic Press.
Yamazaki, H. (1979). *Optik*, *54*, 343.

Yamazaki, S., Kawamoto, H., Saburi, K., & Buchanan, R. (1984). *Scanning Electron Microsc.*, *I*, 23.

Zach, J. (1989). *Optik*, *83*, 30.

Zach, J., & Rose, H. (1988). In *Proc. EUREM 88 (York, UK), vol. 1* (p. 81).

Zhu, X., & Munro, E. (1989). *J. Vac. Sci. Technol. B*, 7, 1862.

Zienkiewicz, O. C., & Cheung, Y. K. (1965). *Engineer*, *220*, 507.

Zurmühl, R. (1965). *Praktische Mathematik für Ingenieure und Physiker* (5th ed.). Berlin: Springer-Verlag.

Zworykin, V. K., Hillier, J., & Snyder, R. L. (1942). *ASTM Bull.*, *117*, 15.

INDEX

S

T

U

V

W

Z

Printed in the United States
By Bookmasters